HUMAN-COMPUTER INTERACTION, 2004, Volume 19, pp. 1–8
Copyright © 2004, Lawrence Erlbaum Associates, Inc.

# Introduction to This Special Issue on Human–Robot Interaction

### Sara Kiesler
*Carnegie Mellon University*

### Pamela Hinds
*Stanford University*

Human–computer interaction (HCI), as a field, has made great strides toward understanding and improving our interactions with computer-based technologies. From the early explorations of direct interaction with computers, we have reached the point where usability, usefulness, and an appreciation of technology's social impact, including its risks, are widely accepted goals in computing. HCI researchers, designers, and usability engineers work in a variety of settings on many kinds of technologies. Recent proceedings of the CHI conference give evidence of this diversity. Topics include not only the office systems where HCI work began, but also tiny mobile devices, Web and Internet services, games, and large networked systems. This special issue introduces a rapidly emerging technology and new focus for HCI—autonomous robots and the human–robot interactions required by these robots.

Until recently, HCI researchers have done little work with robots. Keywords related to robots or to human–robot interaction have not been included

**Sara Kiesler** studies social psychological aspects of technology such as the Internet, networked organizations, and robotics. She is Hillman Professor of Computer Science, Human–Computer Interaction Institute, Carnegie Mellon University, Pennsylvania. **Pamela J. Hinds** studies the impact of technology on individuals and groups. She is an Assistant Professor in the Department of Management Science and Engineering at Stanford University.

in the lists of terms used in human–computer interaction publications or conferences. This state of affairs was reasonable. As Sebastian Thrun's opening article in this special issue explains, today's workhorse robots are mainly programmable industrial machines that offer modest challenges in human–computer interaction. Now, advances in computer technology, artificial intelligence, speech simulation and understanding, and remote controls are leading to breakthroughs in robotic technology that offer significant implications for the HCI interaction community.

Autonomous mobile robots can identify and track a user's position, respond to spoken questions, display text or spatial information, and travel on command while avoiding obstacles. These robots will soon assist in a range of tasks that are unpleasant, unsafe, taxing, confusing, low paid, or boring to people. For example, nurses making rounds in assisted living facilities spend much of their time sorting and administering medications. A robotic assistant could do some of this work, as well as chores that are difficult for elderly people such as fetching newspapers and mail, getting up and down stairs, getting things out of high or low cabinets, and carrying laundry; enabling elderly people to be independent longer. Robotic assistants in the future might act as guards, help fight fires, deliver materials on construction sites and in mines, and distribute goods or help consumers in retail stores. Robots might even provide high-interaction services such as taking blood and coloring hair.

Autonomous robots like these will need to carry out social and intellectual, as well as physical, tasks. Ideally, these robots will create a comfortable experience for people; gain their cooperation; encourage healthy rather than overly dependent behavior in clients, customers, and co-workers; and provide appropriate feedback to remote operators and others involved in the robotic system. Although roboticists are gaining practical experience with mobile, autonomous robots in settings such as museums (Thrun et al., 2000), we lack a principled understanding of how to design robots that will accomplish these more ambitious goals.

## HUMAN–ROBOT INTERACTION IN HCI

HCI, and its sister discipline, human factors, offers a rich resource for research and design in human–robot interaction. Much has been learned in the last 3 decades about how people perceive and think about computer-based technologies, about human constraints on interaction with machines, about the factors that improve usability, and about the primary and secondary effects of technology on people and organizations. Much of this work will be applicable to robots. Nonetheless, autonomous robots are a distinctive case for several reasons.

First, people seem to perceive autonomous robots differently than they do most other computer technologies. People's mental models of autonomous robots are often more anthropomorphic than their models of other systems (Friedman, Kahn, & Hagman, 2003). The tendency for people to anthropomorphize may be fed, in part, by science fiction and, in part, by the powerful impact of autonomous movement on perception (Scholl & Tremoulet, 2000). When we build autonomous robots to look human, we may encourage anthropomorphic mental models of these systems. As Hinds, Roberts, and Jones (this issue) explain, some roboticists argue that humanoid robots provide for a more natural interface than more mechanistic robots. Therefore, humanoid robotics are the focus of much research and development.

A second major reason autonomous robots are a distinctive case in HCI is that robots are ever more likely to be fully mobile, bringing them into physical proximity with other robots, people, and objects. As two articles in this special issue (e.g., Burke, Murphy, Coovert, & Riddle; Yanco, Drury, & Scholtz) make clear, mobile robots will have to negotiate their interactions in a dynamic, sometimes physically challenging, environment. If one or more remote operators partly control the robot, they must help the robot negotiate its interactions in the remote space, creating a complex feedback system. Consider, for example, one such futuristic scenario in a medical setting. We have one such futuristic robot whose task is to sort and dispense medications interacting with an elderly client. At the same time, the robot is designed to sense its clients' state, using indicators such as unusual posture, gestures, or eye movement indicating illness. A remote medical worker monitors this information, adjusting the robot's route or tasks as needed and watching for signs of problems in client states. In this example, the interfaces of interest involve the robot–client, robot–operator, and even multiple person or robot interactions.

A third reason that autonomous robots are a distinctive case for HCI is because these robots make decisions (i.e., they learn about themselves and their world, and they exert at least some control over the information they process and actions they emit). Of course, many computer agents in desktop, automotive, and other computer systems make decisions, and the use of agents is increasing rapidly. Computer agents present interesting HCI issues, for example, to what extent the agent should display confidence intervals for the decisions it makes. An autonomous robotic system will add even more complexity because it must adjust its decisions sensibly and safely to the robot's abilities and to the options available to the robot in a given environment. The system also must detect and respond to changes in the environment and its users. Imagine a robotic walker that guides a frail person and detects when its user is ill or falling or when the environment is dangerous. How much control should such a walker take? How sure of itself should it be? How should it respond if the user wants to turn back, stop, or oppose the robot's suggestions?

As these questions suggest, designing an appropriate interaction scheme and interface for such a system requires an understanding of the people who will use such a system, and of their world. As the ethnography of elders in this special issue shows (Forlizzi, DiSalvo, & Gemperle), designing these robots appropriately will require a deep understanding of the context of use and of the ethical and social considerations surrounding this context.

We do not claim that these problems are entirely new. Design explorations and research in human–robot interaction existed in the field of robotics since at least the mid 1990s. At Interval Research Corporation, for example, Mark Scheef and his colleagues (Scheef, Pinto, Rahardja, Snibbe, & Tow, 2000) built Sparky, a "social robot," and studied children's and adults' reactions to it. Today, many such developments are taking place in Europe and in Japan; for instance, the humanoid Robovie robot described in this special issue (Kanda, Hirano, Eaton, & Ishiguro). MIT's Robotic Life project is an example of design explorations at the edge of robotics and HCI, in which Cynthia Breazeal and her colleagues are trying to create capable robotic creatures with a lifelike presence. Another example in quite a different domain is the work of Brian Scassellati (2000), who builds human-like robots to investigate models of human development. Other domains include space exploration and the military. Over the last few years, research on human–robot interaction has gained increasing attention and funding. The National Science Foundation and the Defense Department's DARPA recently co-sponsored an interdisciplinary workshop in which participants discussed problems of human–robot interaction for Robonaut, a robot that will help astronauts outside a space capsule, and for search and rescue robots (Murphy & Rogers, 2001). Two yearly conferences now provide a forum for articles on human–robot interfaces—the *Humanoid Robots Conference* and the *IEEE RO-MAN Workshop*.

In planning this special issue, we noted that despite the many prior and ongoing activities in robotics related to human–robot interaction, most of the development and the published literature in this area is concerned with technical advances in robotics and computer science that make human–robot interaction possible. Our goal for this issue was to stretch the field of inquiry by focusing especially on behavioral, cognitive, and social aspects of human–robot interaction and the social contexts surrounding human–robot interaction. For example, we hope this special issue will encourage researchers in the field to think about what useful tasks robots can and should do in real social environments, and how to improve how robots communicate and respond to ongoing human communications and behaviors. We invited work that examined human–robot interaction in its social context. We imposed another bias too: Given the comparative absence of systematic empirical investigation in the field, we gave preference to systematic empirical studies and to interdisciplinary collaborations. We also encouraged authors to reflect on the social and ethical issues raised by the deployment of robots in work or everyday life. The HCI community is an especially appropriate place to carry out this kind of analysis because of its legacy of applying behavioral and social science to technical problems and of doing interdisciplinary research and design.

## CONTENTS OF THIS SPECIAL ISSUE

In this special issue of *Human–Computer Interaction,* we present six articles in the emerging area of human–robot interaction.

The first article in this special issue, an invited essay by Sebastian Thrun, provides a technical context for the articles that follow. The author reviews the state of the art in robotics, suggests advances that are likely in the future, and points out some challenges faced in robotics that impinge on human–robot interaction. The author suggests a useful framework for HCI researchers' work in human–robot interaction (i.e., a framework that differentiates among three kinds of robots—industrial robots, professional service robots that will operate in work organizations and public settings, and personal service robots). These three kinds of robots have different capabilities, different user groups, and different contexts of use. This framework will help the HCI community identify some of the greatest opportunities for research in human–robot interaction

The first empirical article in this special issue, by Forlizzi, DiSalvo, and Gemperle, offers a theoretical ecological framework for the design of personal service robots in homes of elderly people. The authors use this framework to show how aging occurs within a local ecology that includes the elder person, the home, products within the home, and important people in the elderly person's life. The authors describe a fascinating ethnography of elders in which they explore how products maintain or lose their usefulness and value for well and ill elders. More generally, the study and the framework should help designers and researchers to consider, and design for, the social context of personal service robots.

The next article in this special issue, by Kanda, Hirano, Eaton, and Ishiguro, presents a field study of two robots that visited a children's elementary school in Japan for 2 weeks, with the purpose of teaching children English. This article is a good example of a field trial with robots. The trial exemplifies the risks and advantages of studying peoples' responses to robots over time in a real social setting. The authors had to understand and cope with problems of a noisy environment and rambunctious children, but they were able to track interactions and the effects of these interactions on learning over time. The children's enthusiasm for the robots waned over the 2-week period, but those children who continued to interact with the robot (mainly those who could understand a bit of the robot's English to begin with), learned from it. Although the effects are modest and the time was short, the results of this study are impressive because this study is the first practical demonstration that students can learn from a humanoid robot.

The third empirical article in this special issue, by Burke, Murphy, Coovert, and Riddle, reports on an opportunistic field study of search and rescue robots used as part of a night rescue training exercise. The authors made careful observations of how remote operators interacted with the robots and one another, and then developed a systematic coding scheme to analyze these interactions. To their surprise, the main human–robot interaction problem was not remote navigation but rather understanding the situation the robot had encountered. The authors describe the interactions among team members who helped the operator understand the state of the robot and the environment. This article is not only an interesting account of the people and robots used in disaster search and rescue, but also points to some of the main human–robot interaction problems in this domain.

The fourth empirical article in this special issue, by Yanco, Drury, and Scholtz, offers a different perspective on HCI for search and rescue robots. The authors took advantage of a yearly robotics IEEE competition for search and rescue robots held to encourage advancements in this field. They developed metrics to compare the usability of the human–robot interface across competitors, and they compared their observations using these metrics with performance scores in the competition. The authors argue that usability standards for other kinds of computer interfaces are only partly applicable to human–robot interfaces. For example, as did the authors of the previous article, these authors conclude that one of the biggest problems in the human–robot interface for search and rescue robots is that the remote operator often lacks accurate situation awareness of the robot's state and the state of the environment in which the robot is located. This problem seems to us to be unique to human–robot interaction, and especially difficult because of simultaneous changes taking place in the operator, the robot, and the robot's environment.

The fifth empirical article in this special issue, by Hinds, Roberts, and Jones, is an experimental laboratory study. The authors explore how people who have to work closely with professional service robots will perceive and work with these robots. This study is one of the first controlled experiments to examine the effect of a humanoid robot appearance on peoples' responses, with a machine-like robot used as a comparison. The study suggests that people may be more willing to share responsibility with humanoid as compared with more machine-like robots, a possibility that has important implications for collaborations in which the robot makes key decisions about the task.

Taken as a whole, these articles represent some of the first systematic empirical research in human–robot interaction. We hope these articles show that human–robot interaction offers many interesting and important problems for the HCI community. More interdisciplinary collaboration between roboticists, behavioral and social scientists, and designers is important, we believe, to advancing the field of human–robot interaction. Roboticists understand the technology and its applications; behavioral scientists and others can provide theory and methods. However, there are plenty of opportunities even for those far from a robotics laboratory. For instance, research on computer agents; avatars; and other ways of representing autonomous, computer-based assistance will contribute to our understanding and design of robots. Useful studies also can proceed using commercial robots such as AIBO and the Help-Mate robot (King & Weiman, 1990), conducting laboratory studies using robot shells and *Wizard of Oz* methods, or performing ethnographic studies of the contexts to which robots may be applied. In general, we see many opportunities for researchers of all stripes and believe that leadership from the HCI community could advance research in human–robot interaction in important ways, influencing the development of the field and the design of robots.

## NOTES

*Support.* We acknowledge the support of the National Science Foundation (IIS-0121426) in preparing this special issue.
*Authors' Present Addresses.* Sara Kiesler, Human–Computer Interaction Institute, Carnegie Mellon University, Pittsburgh, PA 15213. E-mail: kiesler@cs.cmu.edu. Pamela J. Hinds, Management Science & Engineering, Terman 424, Stanford University, Stanford, CA 94305-4026. E-mail: phinds@stanford.edu.

## REFERENCES

Friedman, B., Kahn, P. H., & Hagman, J. (2003). Hardware companions?—What online AIBO discussion forums reveal about the human–robotic relationship. *Proceedings of the CHI 2003 Conference on Human Factors in Computing Systems.* New York: ACM.

King, S., & Weiman, C. (1990). Helpmate autonomous mobile robot navigation system. *Proceedings of the SPIE 1990 Conference on Mobile Robots.* Bellingham, WA: International Society for Optical Engineering.

Murphy, R., & Rogers, E. (2001). *Human–robot interaction: Final report for DARPA/NSF study human–robot interaction.* Retrieved June 5, 2002, from http://www.csc.calpoly.edu/~erogers/HRI/HRI-report-final.html

Scassellati, B. (2000). How robotics and developmental psychology complement each other. *Proceedings of the NSF/DARPA Workshop on Development and Learning.* East Lansing: Michigan State University Press.

Scheef, M., Pinto, J., Rahardja, K., Snibbe, S., & Tow, R. (2000). *Experiences with Sparky, a social robot.* Paper presented at the 2000 Workshop on Interactive Robot Entertainment, Pittsburgh, PA.

Scholl, B. J., & Tremoulet, P. (2000). Perceptual causality and animacy. *Trends in Cognitive Science, 4,* 299–309.

Thrun, S., Beetz, M., Bennewitz, M., Burgard,W., Cremers, A., Dellaert, F., et al. (2000). Probabilistic algorithms and the interactive museum tour-guide robot Minerva. *International Journal of Robotics Research, 19,* 972–999.

---

## ARTICLES IN THIS SPECIAL ISSUE

HUMAN-COMPUTER INTERACTION, 2004, Volume 19, pp. 9–24

# Toward a Framework for Human–Robot Interaction

## Sebastian Thrun
*Stanford University*

### ABSTRACT

The goal of this article is to introduce the reader to the rich and vibrant field of robotics. Robotics is a field in change; the meaning of the term *robot* today differs substantially from the term just 1 decade ago. The primary purpose of this article is to provide a comprehensive description of past- and present-day robotics. It identifies the major epochs of robotic technology and systems—from industrial to service robotics—and characterizes the different styles of human–robot interaction paradigmatic for each epoch. To help set the agenda for research on human–robot interaction, the article articulates some of the most pressing open questions pertaining to modern-day human–robot interaction.

## 1. INTRODUCTION

The field of robotics is changing at an unprecedented pace. At present, most robots operate in industrial settings where they perform tasks such as assembly and transportation. Equipped with minimal sensing and computing, robots are slaved to perform the same repetitive task over and over again. In

**Sebastian Thrun** studies robotics, artificial intelligence, and machine learning; he is an associate professor in the Computer Science Department at Stanford University.

## CONTENTS

the future, robots will provide services directly to people, at our workplaces, and in our homes.

These developments are sparked by a number of contributing factors. Chief among them is an enormous reduction in costs of devices that compute, sense, and actuate. Of no lesser importance are recent advances in robotic autonomy, which have critically enhanced the ability of robotic systems to perform in unstructured and uncertain environments (for an overview, see Thrun, 2002). All these advances have made it possible to develop a new generation of service robots, posed to assist people at work, in their free time, and at home.

From a technological perspective, robotics integrates ideas from information technology with physical embodiment. Robots share with many other physical devices, such as household appliances or cars, the fact that they "inhabit" the same physical spaces as people do in which they manipulate some of the very same objects. As a result, many forms of human–robot interaction involve pointers to spaces or objects that are meaningful to both robots and people (Kortenkamp, Huber, & Bonasso, 1996). Moreover, many robots have to interact directly with people while performing their tasks. This raises the question as to what the right modes are for human robot interaction. What is technologically possible? And what is desirable?

Possibly the biggest difference between robots and other physical devices—such as household appliances—is autonomy. More than any other research discipline, the field of robotics has striven to empower robots with an ability to make their own decisions in broad ranges of situations. Today's most advanced robots can accommodate much broader circumstances than, for example, dishwashers can. Autonomy opens the door to much richer interactions with people: Some researchers consider their systems *social* (Simons et al., 2003) or *sociable* (Breazeal, 2003a). Sociable interaction offers both opportunities and pitfalls. It offers the opportunity for the design of much improved interfaces by exploiting rules and conventions familiar to people in different social contexts. However, sociable interaction does so at the danger of people reflecting capabilities that do not exist into robotic technology. For these and other reasons, it remains unclear if we ever want to interact with robots the same way we interact with our next-door neighbor, our colleagues, or with the people who work in our homes.

## 2. THE THREE KINDS OF ROBOTS

Robotics is a broad discipline. The breadth of the field becomes apparent by contrasting definitions of robots. In 1979, the Robot Institute of America defined a robot as "a reprogrammable, multifunctional manipulator designed to move materials, parts, tools, or specialized devices through various programmed motions for the performance of a variety of tasks" (Russell & Norvig, 1995). In contrast, the Merriam Webster's collegiate dictionary (1993) defines a robot as "An automatic device that performs functions normally ascribed to humans or a machine in the form of a human." A technical introduction into robotic sensors, actuators, and algorithms can be found elsewhere (e.g., Thrun, 2002).

The United Nations (U.N.), in their most recent robotics survey (U.N. and I.F.R.R., 2002), grouped robotics into three major categories. These categories—industrial robotics, professional service robotics, and personal service robotics—are primarily defined through their application domains. These categories also represent different technologies and correspond to different historic phases of robotic development and commercialization.

Industrial robots represent the earliest commercial success, with the most widespread distribution to date. An industrial robot has three essential elements: It manipulates its physical environment (e.g., by picking up a part and placing it somewhere else); it is computer controlled; and it operates in industrial settings, such as on conveyor belts. The boundary between industrial robots and non-robotic manufacturing devices is somewhat fuzzy; the term *robot* is usually used to refer to systems with multiple actuated elements, often arranged in chains (e.g., a robotic arm). Classical applications of industrial robotics include welding, machining, assembly, packaging, palletizing, transportation, and material handling. For example, Figure 1 shows an industrial welding robot in the left panel next to a robotic vehicle for transporting containers on a loading dock in the right panel.

Industrial robotics started in the early 1960s, when the world's first commercial manipulator was sold by Unimate. In the early 1970s, Nissan Corporation automated an entire assembly line with robots, starting a revolution that continues to this day. To date, the vast majority of industrial robots are installed in the automotive industry, where the ratio of human workers to robots is approximately 10:1 (U.N. and I.F.R.R., 2002). The outlook of industrial robots is prosperous. In 2001, the U.N. estimated the operational stock of industrial robots to be 780,600; a number that is expected to grow by just below 25%

until 2005. According to the U.N. (U.N. and I.F.R.R., 2002), the average cost of an industrial robot has decreased by 88.8% between 1990 and 2001. At the same time, U.S. labor costs increased by 50.8%. These opposing trends continue to open up new opportunities for robotic devices to take over jobs previously reserved for human labor. However, industrial robots tend not to interact directly with people. Interface research in this field focuses on techniques for rapidly configuring and programming these robots.

Professional service robots constitute much younger kinds of robots. Service robotics is mostly in its infancy, but the field is growing at a much faster pace than industrial robotics. Just like industrial robots, professional service robots manipulate and navigate their physical environments. However, professional service robots assist people in the pursuit of their professional goals, largely outside industrial settings. Some of these robots operate in environments inaccessible to people, such as robots that clean up nuclear waste (Blackmon et al., 1999; Brady et al., 1998) or navigate abandoned mines (Thrun et al., 2003). Others assist in hospitals, such as the HelpMate® robot (King & Weiman, 1990) shown in Figure 2a, which transports food and medication in hospitals; or the surgical robotic system shown in Figure 2b, used for assisting physicians in surgical procedures. Robot manipulators are also routinely used in chemical and biological labs, where they handle and manipulate substances (e.g., blood samples) with speeds and precisions that people cannot match; recent work has investigated the feasibility of inserting needles into human veins through robotic manipulators (Zivanovic & Davies, 2000). Most professional service applications have emerged in the past decade. According to the U.N. (U.N. and I.F.R.R., 2002), 27% of all operational professional service robots operate underwater, 20% perform demolitions, 15% offer medical services, and 6% serve people in agriculture (e.g., by milking cows; see Reinemann & Smith, 2000). Military applications such as bomb diffusal, search and rescue (Casper, 2002), and support of SWAT teams (Jones, Rock, Burns, & Morris, 2002) are of increasing relevance (U.N. and I.F.R.R., 2002). According to the U.N., the total operational stock of professional service robots in 2001 was 12,400, with a 100% growth expectation by 2005. The amount of direct interaction with people is much larger than in the industrial robotics field, because service robots often share the same physical space with people.

Personal service robots possess the highest expected growth rate. According to optimistic estimates (U.N. and I.F.R.R., 2002), the number of deployed personal service robots will grow from 176,500 in 2001 to 2,021,000 in 2005—a stunning 1,145% increase. Personal service robots assist or entertain people in domestic settings or in recreational activities. Examples include robotic vacuum cleaners, lawn mowers, receptionists, robot assistants to elderly and people with disabilities, wheelchairs, and toys. Figure 3 shows two exam-

*Figure 1.* Industrial robots. (a) A typical welding robot and (b) an autonomous robot for transporting containers on a loading deck (Durrant-Whyte, 1996).

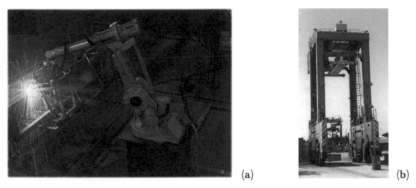

(a)  (b)

*Figure 2.* Professional service robots. (a) The HelpMate hospital delivery robot and (b) a surgical robot by Intuitive Surgical.

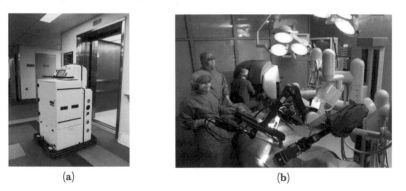

(a)  (b)

*Figure 3.* Personal service robots. (a) The Nursebot, a prototype personal assistant for the elderly. (b) A robotic walker. (c) A robotic vacuum cleaner (Roomba® by iRobot, Inc.).

(a)  (b)  (c)

ples from the medical sector: a robotic assistant to the elderly (Montemerlo, Pineau, Roy, Thrun, & Verma, 2002; Schraft, Schaeffer, & May, 1998) and a robotic walker (Dubowsky, Genot, & Godding, 2002; Glover et al., 2003; Lacey & Dawson-Howe, 1998; Morris et al., 2003). Figure 3c depicts a series of humanoid robots developed with an eye toward domestic use. Another example is shown in Figure 4c: a robotic toy, popular through its use for robotic soccer (Kitano, 1998). Personal service robots are just beginning to emerge. The sales of robotic toys alone are projected to increase by a factor of 10 in the next 4 years (U.N. and I.F.R.R., 2002). Many of these robots must interact with people who, in general, possess no special skills or training to operate a robot. Therefore, finding effective means of interaction is more crucial in this new market segment of robotic technology than in industrial robotics or professional service robotics.

The widely acknowledged shift from industrial to service robotics, and the resulting increase of robots that operate in close proximity to people, raises a number of research and design challenges. Some important challenges are outside the scope of this article, such as those pertaining to safety and cost. From the human–robot interaction perspective, a very important characteristic of these new target domains is that service robots share physical spaces with people. In some applications, these people will be professionals that may be trained to operate robots. In others, they may be children, elderly, or people with disabilities whose ability to adapt to robotic technology may be limited. The design of the interface, although dependent on the specific target application, will require substantial consideration of the end user of the robotic device. Herein lies one of the great challenges that the field of robotics faces today.

## 3. ROBOTIC AUTONOMY

*Autonomy* refers to a robot's ability to accommodate variations in its environment. Different robots exhibit different degrees of autonomy; the degree of autonomy is often measured by relating the degree at which the environment can be varied to the mean time between failures and other factors indicative of robot performance. Human–robot interaction cannot be studied without consideration of a robot's degree of autonomy, because it is a determining factor with regards to the tasks a robot can perform, and the level at which the interaction takes place.

The three kinds of robotics are characterized by different levels of autonomy, largely pertaining to the complexity of environments in which they operate. It should come as little surprise that industrial robots operate at the lowest level of autonomy. In industrial settings, the environment is usually highly engineered to enable robots to perform their tasks in an almost mechanical way.

For example, pick-and-place robots are usually informed of the physical properties of the parts to be manipulated, along with the locations at which to expect parts and where to place them. Driverless transportation vehicles in industrial settings often follow fixed paths defined by guide wires or special paint on the floor. As these examples suggest, careful environment engineering indeed minimizes the amount of autonomy required—a key ingredient of the commercial success of industrial robotics.

This picture is quite different in service robotics. Although environment modifications are still commonplace—the satellite-based global positioning system that helps outdoor robots determine their locations is such a modification—the complexity of service robot environments mandate higher degrees of autonomy than in industrial robotics. The importance of autonomy in service robotics becomes obvious in Figure 5a: This diagram depicts the trajectory of a museum tour guide robot (Burgard et al., 1999) as it toured a crowded museum. Had the museum been empty, the robot would have been able to blindly follow the same trajectory over and over again—just as industrial robots tend to repeatedly execute the same sequence of actions. The unpredictable behavior of the museum visitors, however, forced the robot to adopt detours. The ability to do so sets this robot apart from many industrial applications.

Autonomy enabling technology has been a core focus of robotics research in the past decade. One branch of research is concerned with acquiring environmental models. An example is shown in Figure 5b, which depicts a two-dimensional (2D) map of a nursing home environment acquired by the robot in Figure 3b by way of its laser range finders. Such a 2D map is only a projection of the true 3D environment; nevertheless, paired with a planning system, it is sufficiently rich to enable the robot to navigate in the absence of environmental modifications. Other research has focused on the capability to detect and accommodate people. In general, robots that operate in close proximity to people require a high degree of autonomy, partially because of safety concerns and partially because people are less predictable than most objects. It is common practice to endow service robots with sensors capable of detecting and tracking people (Schulz, Burgard, Fox, & Cremers, 2001). Some researchers have gone as far as devising techniques whereby robots learn about people's routine behavior and actively step out of the way when people approach (Bennewitz, Burgard, & Thrun, 2003).

The type and degree of autonomy in service robotics varies more with the specific tasks a robot is asked to perform and the environment in which it operates. Personal robots tend to be etching at low-cost markets. As a result, endowing a personal robot with autonomy can be significantly more difficult than its more expensive professional relative. For example, the robotic dog shown in Figure 4 is equipped with a low-resolution CCD camera and an

*Figure 4.* Faces in robotics. (a) Animated face on the robot Grace (Simons et al., 2003), (b) mechatronic face of KISMET (Breazeal, 2003a), and (c) the Sony AIBO® robotic dog.

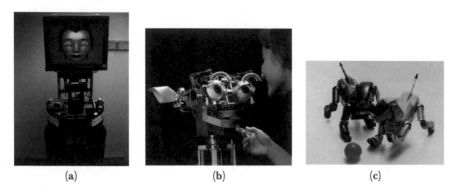

(a)　　　　　　　　　(b)　　　　　　　　　(c)

*Figure 5.* (a) Path taken by an autonomous service robot, and (b) two-dimensional map "learned" by a robot.

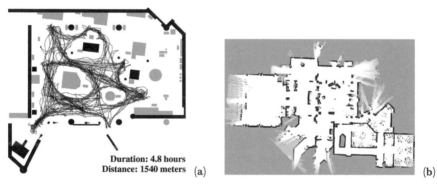

Duration: 4.8 hours
Distance: 1540 meters (a)

(b)

*Figure 6.* (a) Humanoid robots ASIMO and P3 by Honda, and (b) Tour guide robot Minerva, with an actuated humanoid face but a nonhumanoid torso.

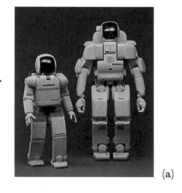

(a)　　　　　　　　　(b)

onboard computer whose processing power lags behind most professional service robots by orders of magnitude—which adds to the challenge of making it
autonomous.

## 4. HUMAN–ROBOT INTERFACES

Robots, like most other technological artifacts, require user interfaces for
interacting with people. Interfaces for industrial robots tend to differ from interfaces for professional service robots, which in turn differ from that for personal service robots. In industrial robotics, the opportunity for human–robot
interaction is limited, because industrial robots tend not to interact directly
with people. Instead, their operational space is usually strictly separated from
that of human workers. Interface technology in industrial robotics is largely re-
stricted to special purpose programming languages and graphical simulation
tools (Nof, 1999), which have become indispensable for configuring robotic
manipulators. Some researchers have developed techniques for programming
robots through demonstration (Friedrich, Münch, Dillman, Bocionek, &
Sassin, 1996; Ikeuchi, Suehiro, & Kang, 1996; Mataric, 1994; Schaal, 1997).
The idea here is that a human demonstrates a task (e.g., an assembly task)
while being monitored by a robot. From that demonstration, the robotic device learns a strategy for performing the same tasks by itself.

Most service robots will require richer interfaces. Here, I distinguish interfaces for indirect interaction with interfaces for direct interaction. For the sake
of this article, I define indirect interaction to be the interaction that takes place
when a person operates a robot; for example, an operator gives a command
that the robot then executes. A robotic surgeon interacts indirectly with a surgical robot, as the one shown in Figure 2b, in that the robot merely amplifies
the surgeon's force. Direct interaction is different in that the robot acts on its
own; the robot acts and the person responds or vice versa.

A nice way to distinguish indirect from direct interaction pertains to the
flow of information and control: In indirect interaction, the operator commands the robot, which communicates back to the operator information about
its environment, its task, and its behavior. In direct interaction, the information flow is bidirectional: Information is communicated between the robot
and people in both directions, and the robot and the person are interacting on
"equal footing." An example is the robotic caregiver in Figure 3a, which interacts with people in ways motivated by people's interactions with nurses. In
particular, it asks questions, and it can also respond to questions asked by people. As a general rule of thumb, the interaction with professional service robots
is usually indirect, whereas the interaction with personal service robots tends
to be more direct. There are exceptions to this rule, such as the robotic vacuum

cleaner in Figure 3c, which is a personal service robot whose interaction is entirely indirect.

There exists a range of interface technologies for indirect interaction. The classical interface is the master–slave interface in which a robot exactly duplicates the same physical motion of its operator. A recent implementation of this idea is given by Robonaut (Ambrose et al., 2001), a robot developed as a telepresence device on a space station. The goal of this project is to demonstrate that a robotic system can perform repairs and inspections of space flight hardware originally designed for human servicing. Some robots are operated remotely using interfaces familiar from radio controlled cars (Casper, 2002); others possess haptic displays and control interfaces (Ruspini, Kolarov, & Khatib, 1997).

In service robotics, the utility of direct interaction is much less established than that of indirect interaction. To study direct interaction, numerous research prototypes have been equipped with speech synthesizers and recognizers or sound-synthesizing devices. Some robots only generate speech but do not understand spoken language (Thrun et al., 2000); others also understand spoken language (Asoh et al., 1997; Bischoff & Graefe, 2003) or use keyboard interfaces to bypass speech recognition altogether (Torrance, 1994). Speech as output modality is easy to control and can be quite effective.

Several researchers have reported excellent results for speech understanding in office robots in which speakers were instructed with regards to vocabulary the robot was able to understand (Asoh et al., 1997). Encouraging results have also been reported for a museum tour guide robot that understands spoken commands in multiple languages (Bischoff & Graefe, 2003). To my knowledge, none of these systems have been evaluated systematically with regards to the effectiveness of the speech interface. Roy, Pineau, and Thrun (2000) studied a service robot in an elderly care facility in which participants were not instructed about the robot's vocabulary. The authors found that the speech interface was difficult to use. Only about 10% of the words used by the target group were in the robot's vocabulary. Misunderstanding may be exacerbated if the robot's ability to talk creates a false perception of human-level intelligence (Nourbakhsh, Rosenberg, & Thrun, 1999; Schulte et al., 1999).

A number of robots carry graphical screens capable of displaying information to the user (Nourbakhsh et al., 1999; Simons et al., 2003). Researchers have used both regular and touch-sensitive displays (Nourbakhsh et al., 1999). The information on the display may be organized in menus similar to the ones found on typical information kiosks. Some researchers have used the display to project an animated face (Simons et al., 2003), such as the one shown in Figure 4a. Gesture recognition (Kahn, Swain, Prokopowicz, & Firby, 1996; Kortenkamp et al., 1996; Perzanowski, Adams, Schultz, & Marsh, 2000; Waldherr, Thrun, & Romero, 2000) has also been investigated, as has gaze tracking (Heinzmann & Zelinsky, 1998; Zelinsky & Heinzmann, 1996), as an interface to refer to physical objects in a robot's workspace. Such interfaces have shown moderate promise for robots assisting people with severe disabilities. A recent study has investigated modalities as diverse as head motion, breath expulsion, and electrooculographic signals (eye motion recorded by measuring electrical activity in the vicinity of the eye) as alternative interfaces for people with disabilities (Mazo et al., 2000).

Interface technologies also exist that are unique to robotics, in that they require physical embodiment. A classic example is that of a mechatronic head (Breazeal, 2003a, 2003b) shown in Figure 4b. The face is capable of exhibiting different facial expressions, such as a smile, a frown, and a surprised look. The face does this by moving actuated elements into position reminiscent of human muscular movement when expressing certain emotions. In the past decade, dozens of such faces have been developed for service robot applications, with varying degrees of dexterity and expressiveness. Many robotic faces are able to change the expression of the mouth and the eyes, emulating basic expressions such as smiling and frowning. The face shown in Figure 4b possesses 15 independently actuated elements and, consequently, can express quite a range of different postures.

Some researchers have begun to explore the social aspects of service robotics. Humanoid robots, by virtue of their appearance and behavior, may appeal to people differently than other technological artifacts, as a recent survey suggested (Fong, Nourbakhsh, & Dautenhahn, 2003). Most research thus far on sociable robots has focused on humanoid robots and robots with humanoid elements (see Figure 6a). For instance, Kiesler and Goetz (2002) reported on experiments in which the presence and absence of humanoid features and the behavior of the robot influenced people's assumptions about its capabilities and social inclinations. Scassellati (2000) investigated the use of humanoid robots to understand human development. His prototype robot uses its ability to track where a toddler is looking to create a model of gaze movement and focus of attention. Robots like his may be used in the future to help train autistic children to behave more socially. We do not yet know if children will interact with a robot as they will with a person. More generally, it remains unclear whether we, the people who will ultimately interact with service robots on a daily basis, will seek social-style interactions with robots that parallel our interactions with other people.

## 5. OPEN QUESTIONS

Human–robot interaction is a field in change. The technical developments in robotics have been so fast that the types of interactions that can be performed today differ substantially from those that were possible even 1 decade ago. Interaction can only be studied relative to the available technology. Because robotics is still in its infancy, so is the field of human–robot interaction. What follows are several illustrative open questions that occurred to me while writing this article:

- How effective is a mechatronic face over a simulated one? How effective is a humanoid torso over a nonhumanoid one? Does the physical shape of a robot affect the way people interact with it?
- What happens when robots work with or interact with groups of people? How do interfaces scale up to multiple "operators"? Can we build robots that recognize and interact directly with groups?
- To what extent is progress in human–robot interaction tied to progress in other core disciplines of robotics, such as autonomy? In which way will future advances in robotic autonomy change the way we interact with robots?
- To what extent does robotics benefit from direct interactions, as opposed to the well-established indirect interaction? Assuming that interaction is merely a means to an end, does direct interaction improve the on-task performance relative to indirect interaction? Does it make robots amenable to a broader range of problem domains (e.g., elderly people unfamiliar with robotic technology)? How will these interfaces be affected when service robots are as familiar to people as personal computers are today?
- Much of the recent commercial success of PDAs can be attributed to a specialized alphabet—graffiti—designed to maximize the recognition accuracy. Will there be a graffiti for robotics? If so, what will it look like, and what will it be good for?
- When referencing an object, is it better to recognize gestures of a person pointing toward the object, or is it more effective to have a person use a touch-sensitive display of a camera image containing the object?

Many researchers in robotics now recognize that human–robot interaction plays a pivotal role in personal service robotics (Rogers & Murphy, 2001). What constitutes an appropriate interface, however, is subject to much debate. Whatever the answer, it is likely to change in the future. Some of these changes will arise from the continuing stream of advances in robotic autonomy, which inevitably shifts the focus of the interaction to increasingly higher levels. Some changes will arise as a result of changed expectations: As we get used to robots in our environments, we might develop interaction skills that are noticeably different from our interactions with other technological artifacts, or with people. Regardless of the specifics of these developments, I conjecture that human–robot interaction will soon become a primary concern in the majority of robotic applications.

## NOTES

*Acknowledgments.* I acknowledge insightful comments by Pamela Hinds and Sara Kiesler on earlier versions of this article.

*Support.* This research is sponsored by DARPA's MARS Program (Contracts N66001-01-C-6018 and NBCH1020014) and the National Science Foundation (CAREER Grant IIS-9876136 and regular grant IIS-9877033), all of which is gratefully acknowledged.

*Author's Present Address.* Sebastian Thrun, Department of Computer Science, Gates Building, Room 154, Stanford University, Stanford, CA 94305. E-mail: thrun@ stanford.edu.

*HCI Editorial Record.* First manuscript received September 3, 2003. Accepted by Sara Kiesler and Pamela Hinds. Final manuscript received October 5, 2003. — *Editor*

## REFERENCES

Ambrose, R., Askew, R., Bluethmann, W., Diftler, M., Goza, S., Magruder, D., et al. (2001). The development of the robonaut system for space operations. *Proceedings of ICAR 2001, Invited Session on Space Robotics.*

Asoh, H., Hayamizu, S., Isao, H., Motomura, Y., Akaho, S., & Matsui, T. (1997). Socially embedded learning of office-conversant robot jijo-2. *Proceedings of the IJCAI 97 Conference.* Menlo Park, CA: AAAI.

Bennewitz, M., Burgard, W., & Thrun, S. (2003). Adapting navigation strategies using motion patterns of people. *Proceedings of the IEEE International Conference on Robotics and Automation (ICRA).* Piscataway, NJ: IEEE.

Bischoff, R., & Graefe, V. (2003). Hermes: An intelligent humanoid robot, designed and tested for dependability. *Proceedings of the International Symposium on Experimental Robotics (ISER).* Berlin: Springer Verlag.

Blackmon, T., Thayer, S., Teza, J., Broz, V., Osborn, J., Hebert, M., et al. (1999). Virtual reality mapping system for Chernobyl accident site assessment. *Proceedings of the SPIE.* Bellingham, WA: SPIE Press.

Brady, K., Tarn, T., Xi, N., Love, L., Lloyd, P., Davis, H., et al. (1998). Remote systems for waste retrieval from the Oak Ridge national laboratory gunite tanks. *International Journal of Robotics Research, 17,* 450–460.

Breazeal, C. (2003a). Emotion and sociable humanoid robots. *International Journal of Human–Computer Studies, 59,* 119–155.

Breazeal, C. (2003b). Towards sociable robots. *Robotics and Autonomous Systems, 42,* 167–175.

Burgard, W., Cremers, A., Fox, D., Hähnel, D., Lakemeyer, G., Schulz, D., et al. (1999). Experiences with an interactive museum tour-guide robot. *Artificial Intelligence, 114,* 3–55.

Casper, J. (2002). *Human–robot interactions during the robot-assisted urban search and rescue response at the World Trade Center.* Unpublished master's thesis, Computer Science and Engineering, University of South Florida, Tampa.

Dubowsky, S., Genot, F., & Godding, S. (2002). PAMM: A robotic aid to the elderly for mobility assistance and monitoring: A "helping-hand" for the elderly. *Proceedings of the IEEE International Conference on Robotics and Automation (ICRA).* Piscataway, NJ: IEEE.

Durrant-Whyte, H. (1996). Autonomous guided vehicle for cargo handling applications. *International Journal of Robotics Research, 15.*

Fong, T., Nourbakhsh, I., & Dautenhahn, K. (2003). A survey of socially interactive robots. *Robotics and Autonomous Systems, 42,* 143–166.

Friedrich, H., Münch, S., Dillman, R., Bocionek, S., & Sassin, M. (1996). Robot programming by demonstration (rpd): Supporting the induction by human interaction. *Machine Learning, 23*(2/3), 5–46.

Glover, J., Holstius, D., Manojlovich, M., Montgomery, K., Powers, A., Wu, J., et al. (2003). *A robotically-augmented walker for older adults* (Technical Report CMU–CS–03–170). Pittsburgh, PA: Carnegie Mellon University, Computer Science Department.

Heinzmann, J., & Zelinsky, A. (1998). 3-D facial pose and gaze point estimation using a robust real-time tracking paradigm. *Proceedings of the IEEE International Conference on Automatic Face and Gesture Recognition.* Piscataway, NJ: IEEE.

Ikeuchi, K., Suehiro, T., & Kang, S. (1996). Assembly plan from observation. In K. Ikeuchi & M. Veloso (Eds.), *Symbolic visual learning* (pp. 193–224). New York: Oxford University Press.

Jones, H., Rock, S., Burns, D., & Morris, S. (2002). Autonomous robots in SWAT applications: Research, design, and operations challenges. *Proceedings of the 2002 Symposium for the Association of Unmanned Vehicle Systems International (AUVSI '02).* Orlando, FL: AUVSI.

Kahn, R., Swain, M., Prokopowicz, P., & Firby, R. (1996). Gesture recognition using the Perseus architecture. *Proceedings of the IEEE Conference on Computer Vision and Pattern Recognition.* Piscataway, NJ: IEEE.

Kiesler, S., & Goetz, J. (2002). Mental models of robotic assistants. *Proceedings of the CHI 2002 Conference on Human Factors in Computing Systems.* New York: ACM.

King, S., & Weiman, C. (1990). Helpmate autonomous mobile robot navigation system. *Proceedings of the SPIE 1990 Conference on Mobile Robots.* Piscataway, NJ: IEEE.

Kitano, H. (Ed.). (1998). *Proceedings of RoboCup 97: The First Robot World Cup Soccer Games and Conferences.* Berlin: Springer-Verlag.

Kortenkamp, D., Huber, E., & Bonasso, P. (1996). Recognizing and interpreting gestures on a mobile robot. *Proceedings of the AAAI 96 National Conference on Artificial Intelligence.* Menlo Park, CA: AAAI/MIT Press.

Lacey, G., & Dawson-Howe, K. (1998). The application of robotics to a mobility aid for the elderly blind. *Robotics and Autonomous Systems, 23,* 245–252.

Mataric, M. (1994). Learning motor skills by imitation. *Proceedings of the AAAI 98 Spring Symposium.* Menlo Park, CA: AAAI.

Mazo, M., Garcia, J. C., Rodriguez, F. J., Urena, J., Lazaro, J. L., & Espinosa, F. (2000). Integral system for assisted mobility. *IEEE Robotics and Automation Magazine, 129*(1–4), 1–15.

*Merriam Webster's collegiate dictionary* (10th ed.). (1993). Springfield, MA: Merriam Webster, Inc.

Montemerlo, M., Pineau, J., Roy, N., Thrun, S., & Verma, V. (2002). Experiences with a mobile robotic guide for the elderly. *Proceedings of the AAAI 2002 National Conference on Artificial Intelligence.* Menlo Park, CA: AAAI.

Morris, A., Donamukkala, R., Kapuria, A., Steinfeld, A., Matthews, J., Dunbar-Jacobs, J., et al. (2003). A robotic walker that provides guidance. *Proceedings of the IEEE International Conference on Robotics and Automation (ICRA).* Piscataway, NJ: IEEE.

Nof, S. (Ed.). (1999). *Handbook of industrial robotics* (2nd ed.). New York, NY: Wiley.

Nourbakhsh, I., Bobenage, J., Grange, S., Lutz, R., Meyer, R., & Soto, A. (1999). An affective mobile robot with a full-time job. *Artificial Intelligence, 114,* 95–124.

Perzanowski, D., Adams, W., Schultz, A., & Marsh, E. (2000). Towards seamless integration in a multi-modal interface. *Proceedings of the Workshop on Interactive Robotics and Entertainment (WIRE).* Pittsburgh, PA: Carnegie Mellon University.

Reinemann, D., & Smith, D. (2000). Evaluation of automatic milking systems for the United States. In H. Hogeveen & A. Meijering (Eds.), *Robotic milking, Proceedings of the International Symposium.* The Netherlands: Lelystad.

Rogers, E., & Murphy, R. (2001). *Final report for DARPA/NSF study on human–robot interaction.* Retrieved April 1, 2004, from http://www.aic.nrl.navy.mil/hri/nsfdarpa/index.html

Roy, N., Pineau, J., & Thrun, S. (2000). Spoken dialogue management using probabilistic reasoning. *Proceedings of the 38th Annual Meeting of the Association for Computational Linguistics (ACL-2000).* San Francisco, CA: Morgan Kaufmann Pub.

Ruspini, D., Kolarov, K., & Khatib, O. (1997). The haptic display of complex graphical environments. *Proceedings of the SIGGRAPH 97 Conference on Computer Graphics.* New York, NY: ACM SIGGRAPH.

Russell, S., & Norvig, P. (1995). *Artificial intelligence: A modern approach.* Englewood Cliffs, NJ: Prentice Hall.

Scassellati, B. (2000). How robotics and developmental psychology complement each other. *Proceedings of the NSF/DARPA 2000 Workshop on Development and Learning.* Lansing: Michigan State University Press.

Schaal, S. (1997). Learning from demonstration. In M. C. Mozer, M. Jordan, & T. Petshe (Eds.), *Advances in neural information processing systems* (Vol. 9, pp. 1040–1046). Cambridge, MA: MIT Press.

Schraft, R. D., Schaeffer, C., & May, T. (1998). Care-o-bot: The concept of a system for assisting elderly or disabled persons in home environments. *IECON: Proceedings of the IEEE 24th Annual Conference (Vol. 4).* Piscataway, NJ: IEEE.

Schulte, J., Rosenberg, C., & Thrun, S. (1999). Spontaneous short-term interaction with mobile robots in public places. *Proceedings of the IEEE International Conference on Robotics and Automation (ICRA).* Piscataway, NJ: IEEE.

Schulz, D., Burgard, W., Fox, D., & Cremers, A. (2001). Tracking multiple moving targets with a mobile robot using particles filters and statistical data association. *Proceedings of the IEEE International Conference on Robotics and Automation.* Piscataway, NJ: IEEE.

Simons, R., Goldberg, D., Goode, A., Montemerlo, M., Roy, N., Sellner, B., et al. (2003, Summer). GRACE: An autonomous robot for the AAAI robot challenge. *AI Magazine, 24*(2), 51–72.

Thrun, S. (2002). Robotics. In S. Russell & P. Norvig (Eds.), *Artificial intelligence: A modern approach* (2nd ed., pp. 901–946). Englewood Cliffs, NJ: Prentice Hall.

Thrun, S., Beetz, M., Bennewitz, M., Burgard,W., Cremers, A., Dellaert, F., et al. (2000). Probabilistic algorithms and the interactive museum tour-guide robot Minerva. *International Journal of Robotics Research, 19,* 972–999.

Thrun, S., Hähnel, D., Ferguson, D., Montemerlo, M., Triebel, R., Burgard, W., et al. (2003). A system for volumetric robotic mapping of abandoned mines. *Proceedings of the IEEE International Conference on Robotics and Automation (ICRA)*. Piscataway, NJ: IEEE.

Torrance, M. C. (1994). *Natural communication with robots.* Unpublished master's thesis, MIT Department of Electrical Engineering and Computer Science, Cambridge, MA.

U.N. and I.F.R.R. (2002). *United Nations and The International Federation of Robotics: World Robotics 2002.* New York and Geneva: United Nations.

Waldherr, S., Thrun, S., & Romero, R. (2000). A gesture-based interface for human–robot interaction. *Autonomous Robots, 9,* 151–173.

Zelinsky, A., & Heinzmann, J. (1996). Human–robot interaction using facial gesture recognition. *Proceedings of the IEEE International Workshop on Robot and Human Communication*. Silver Springs, MD: IEEE.

Zivanovic, A., & Davies, B. (2000). A robotic system for blood sampling. *IEEE Transactions on Information Technology in Biomedicine, 4*(1).

HUMAN-COMPUTER INTERACTION, 2004, Volume 19, pp. 25–59

# Assistive Robotics and an Ecology of Elders Living Independently in Their Homes

Jodi Forlizzi, Carl DiSalvo, and
Francine Gemperle
*Carnegie Mellon University*

## ABSTRACT

For elders who remain independent in their homes, the home becomes more than just a place to eat and sleep. The home becomes a place where people care for each other, and it gradually subsumes all activities. This article reports on an ethnographic study of aging adults who live independently in their homes. Seventeen elders aged 60 through 90 were interviewed and observed in their homes in 2 Midwestern cities. The goal is to understand how robotic products might assist these people, helping them to stay independent and active longer. The experience of aging is described as an *ecology of aging* made up of people, products, and activities taking place in a local environment of the home and the surrounding community. In this environment, product successes and failures often have a dramatic impact on the ecology, throwing off a delicate balance.

**Jodi Forlizzi** is an interaction designer with an interest in the intersection of aesthetic, assistive, and social products; she is an Assistant Professor in the Human-Computer Interaction Institute and the School of Design at Carnegie Mellon University. **Carl DiSalvo** is a designer with an interest in the relation among agency, the body, power, and design; he is a PhD candidate in the School of Design at Carnegie Mellon University. **Francine Gemperle** is an industrial designer with an interest in the human body; she is a Special Research Faculty in the Institute for Complex Engineered Systems at Carnegie Mellon University.

**CONTENTS**

When a breakdown occurs, family members and other caregivers have to intervene, threatening elders' independence and identity. This article highlights the interest in how the elder ecology can be supported by new robotic products that are conceived of as a part of this interdependent system. It is recommended that the design of these products fit the ecology as part of the system, support elders' values, and adapt to all of the members of the ecology who will interact with them.

# 1. INTRODUCTION

The United States is currently witnessing a rapid increase in the number of elderly people. The U.S. Census (2000) estimated that there will be about 12 million people over age 85 in 2040. Many are expected to need physical and cognitive assistance. Nursing homes and other care facilities can provide some assistance, but they already suffer from space and staff shortages. As the elder population continues to grow, the greatest need for assistance will come from elders who live independently in their own homes.

Advanced technology—artificial intelligence, image processing, and speech simulation and speech processing—can be used to help elders and caregivers. To date, many of the breakthroughs in advanced and autonomous technology, such as assistive robotics, have only begun to offer social and economic benefits (Computing Research Association, 2003). These systems are the precursor to tomorrow's assistive robotic products. Even in this early stage, these systems should address the social, emotional, and environmental needs of elders and caregivers. If inappropriately designed, assistive products come at a great cost to society because they will not help those they were designed for and because they will be unused.

Our goal is to aid the design and development of assistive robotic products for the elderly through the development of design guidelines grounded in ethnographic research. In particular, we are interested in supporting elders who wish to remain independent in their homes. Elders are healthier and report a higher quality of life when they are able to stay in their own homes (Lawton, 1982). There are also substantial financial incentives to helping elders live at home rather than in dedicated care facilities. The questions we need to address are whether robotic products are appropriate assistive products to support elders; and if so, how to introduce them into the context of the home.

We define a product, after design historian Victor Margolin (1997), as any of a broad range of artifacts, services, and environments in the world. A future robotic product might be an intelligent information appliance or a system of distributed products and services within one's home. We believe assistive robotic products will exist in elders' homes along with their treasured possessions, mundane appliances, and a variety of assistive products. We aim to understand how robotic products should fit within elders' home environments. A future step will be to assess specific product ideas—appropriate tasks for the product, what form it might take, what materials it might be constructed from, what senses it might engage, and what interactions it might offer. Together, these design decisions will set the right expectations about how a robotic product might be used.

This article reports an ethnography of the daily domestic lives of elders, their living environments, the products and services embedded in those environments, and the activities that elders engage in. We believe that by better understanding elders' current relations to people, products, and activities, we can begin to ask appropriate questions about how future robotic products might fit within the home.

## 1.1. Context of Assistive Robotics

Technological advances are currently being directed to assist the elder population. These products, which can be charted in the assistive robotics literature, emphasize the independence of the elderly as a primary goal. They provide support for a range of basic activities, including eating, bathing, dressing, and toileting (MOVAID, 2002; RAID Workstation, 2002; RAIL, 2002). They support mobility in the form of ambulation assistance and rehabilitation (GuideCane, 2002; Haptica, 2002; Morris et al., 2002; NavChair, 2002; Wheelesley, 2002). They provide household maintenance in the form of monitoring and maintaining safety in particular environments (Mynatt, Essa, & Rogers, 2000). However, many of these products have been designed with little consideration of the social, aesthetic, and emotional relations that elders will form with the product. Future assistive robotic products will need to move beyond task-based interactions and be attractive, affordable, and nonstigmatizing. Accessibility, ease of use, and reliability will also be particularly important for aging users.

An example of how people's values can affect their usage of an assistive product may be seen in the case of current mobility aids. Walkers, rollators, and canes assist in mobility for all who use them. For the elder population specifically, these products mediate the activities of daily living, provide opportunities for partaking in social activities, and reduce the risk of falls. Studies of elders have shown that nearly one third of these devices are abandoned within the first 3 months (Guralnik et al., 1993); the disuse rate is as high as 54% (Scherer & Galvin, 1994). These products' appearance inhibits many from using them in normal social situations (Hirsch et al., 2000). Elders who do use walkers are inspired by the autonomy they afford; for these people, aesthetic considerations are secondary (Mann, Hurren, Tomita, & Charvat, 1995).

Roboticists have developed a number of mobility enhancing assistive technologies. Most of these share control over motion with the user, undertaking path navigation and obstacle avoidance (Lankenau & Röfer, 2001; Mazo, 2001; Prassler, Scholz, & Fiorini, 2001). The GUIDO (2002) walker provides power-assisted wall or corridor following. The PAMM (Dubowsky et al., 2000) project focuses on navigation for residents in an eldercare facility, adopting a customized holonomic walker frame as its physical form. Although these examples are inspirational, we need to better understand the context that these products will be used in.

Our examination of the emerging field of rehabilitative robotics revealed several opportunities for design research. The home is a growing area of need. Many debilitating accidents happen to the elderly and infirm while unattended at home (Living at Home, 2002). As robotic products emerge to address these safety problems, design research can support their broader usefulness and desirability to elders in their own environments.

## 1.2. Ethnographic Studies of Aging

The social sciences have a rich body of literature on aging and the elderly, much of it having implications for product design. Within sociology and anthropology, substantial ethnographies have examined the individual's experience of aging and interpreted that experience within social, cultural, and even economic frameworks (Bailey, 1986; Golant, 1984; Hazan, 1994; Silverman, 1987; Ward, La Gory, & Sherman, 1988). These works provide valuable insight into understanding the role of objects, environments, and activities within the lives of elders and caregivers; they give designers clues for future product designs.

We found two of these studies to be of particular interest. The "Casser Maison" Ritual (Marcoux, 2001) provides a material culture perspective on aging. The author followed an elder who downsized her home and redistributed her possessions to others in preparation for moving to a care facility. In this situation, the elder retained products whose functions served her particular impairments as well as those that signified her former lifestyle. Dorfman (1994) described an "ideal" aging experience in an extensive ethnography of elders in a Quaker community. The context for the study, an upscale residential retirement facility, offered an environment Dorfman felt was void of many of the stresses common to elders. The values of remaining autonomous, sustaining personal growth, helping others, maintaining social ties, and experiencing pleasures were identified as important for this age group (Dorfman, 1994).

Our investigations identified a clear need for design research focused on how robotic products might support or hinder the values identified in this previous research; for example, keeping possessions that link one to the significant past. We focused our research on elders' relations with products they currently own. In an earlier study, we found that assistive products can feel threatening as well as helpful to people. One woman refused to install bathroom grab bars although her husband had fallen several times in the bathroom; the bars would have ruined the decor of the house (Hirsch et al., 2000). In a related study for a Pittsburgh company developing a wearable monitoring product to improve health and fitness practices, we asked men and women, ages 30 to 55, to describe their reactions to a sleek, stylish, arm-worn medical monitoring device (McCormack & Forlizzi, 2000). Despite the fashionable product form, over one half of the participants said they would not feel comfortable wearing it, particularly at work and in other public places. One third of the participants described the product as reminding them of a lie detector or a blood sugar monitor, despite the high design of the product form. These preliminary studies suggested that we need a better understanding of how assistive products will be perceived and used.

## 1.3. Ecology of Aging

Our literature review and our previous research led to the development of a theoretical framework to guide our ethnographic research. An *ecology* is a set of interdependent parts that have particular relations within a system. Whether one is studying migration patterns in Liberia or the use of cleaning products in California homes, an ecological framework can be useful to examine relations among components of people's experience. In anthropology, *cultural ecology* is the study of the symbiotic relation between people and their social environment (Netting, 1986). Researchers using a cultural ecology approach collect detailed descriptions of people's behavior in their daily environments (Harris, 1979).

More recently, Nardi and O'Day (1999) used the term *information ecology* to describe an interrelated system of people, practices, values, and technologies within a local environment. An information ecology is used to situate new technologies ethically and responsibly, and to understand technology as a catalyst for change. Bell (2002) used the term *ecology* even more broadly, to include all the aspects of a specific experience in context. According to Bell (2001), cultural ecologies and the ethnographic research behind them help to

convey an experience, a sense, a glimpse, or a window into another world ... a way of talking about deep cultural patterns that implicate everything we do. Knowing these stories, interests, and patterns makes it possible to design and develop products and services that fit (intuitively) into people's lives. (p. 2)

Bell's (2001) approach seems most relevant for new product design because it offers a mechanism for examining multifaceted aspects of products.

We offer our ecological framework as a way to make sense of the experience of aging. It helps us to sift through the complex interactions between people, products, and activities, and the experiences that result. These interactions take place in a local environment bounded roughly by the home and the elder community. We have found it useful to consider these interconnected components of experience as an ecology of aging.

The components of the ecology of aging include people, products, the built environment, and the community. The components themselves can be systems or networks. For example, the elder's community is also a social network. Components may or may not reflect the roles and functionality they have in the rest of the society. The nurse's approach to providing care is drastically different from that of the superintendent in an apartment building, but both may be called on to assist with a caregiving task.

The components of the ecology of aging are part of a system and are interconnected in the following ways: First, they are adaptive. If one part of the elder's life breaks down (e.g., an elder is no longer able to drive safely), another part must change (the elder will rely on family and make less frequent trips, or hire a community taxi service created specially for seniors). Second, the flow of information among components can be complex and can have unexpected consequences. An elder may rely on his or her primary care physician for information ranging from blood pressure to how to deal with depression. The physician, in turn, may rely on the elder's family for reports on the elder's general physical and mental well-being. The elder, in turn, may feel the family is interfering, and tell the family less. Third, the components are dynamic and evolving. An elder who has broken a hip will have a myriad of opportunities for assistance in the first few critical weeks, including family, institutional care, home health care, and private and government services. Choosing any one of these can cause the particular experience of an independent elder to change greatly. Finally, the ecology has the potential to break down. Continuing with the aforementioned example, if an elder's family imposes a move to a care institution, the outcome may be more detrimental than beneficial to the elder, resulting in reduced quality of life and well-being.

Our focus in this study was specifically on the role of products within the ecology of aging. Products play a role in a balanced ecology. They help well elders in a variety of activities and experiences, and support independence and well-being. For example, Mrs. G. disseminated information about social events at her community center, and made sure newcomers felt welcome by telling jokes and giving small gifts. Mrs. C. befriended her cleaning lady, preparing a home-cooked meal to share on housecleaning days. Figure 1a depicts an elder within a healthy ecology of family and social connections interacting with products and undertaking activities, connected and vital within a local environment.

The ecology of an elder experiencing decline can be described as one of imbalance. Changes in physical and cognitive abilities contribute to fundamental changes in product interactions. In turn, the elder is less able to undertake activities, and may relinquish independence and rely on assistance. For example, Mrs. R. was clearly struggling to manage her household, and was hurt and upset that her son had begun to "help" by removing items such as her prized Victorola. Mrs. L. relied on a local meal delivery service, but did not like the way the food was prepared and had begun to lose substantial weight.

Figure 1b depicts the ecology of a declining elder who can no longer use all the products she formerly relied on. As a result, a gap is created between the elder and her environment, and a contraction of physical and social lifespace occurs.

*Figure 1.* (a) A balanced elder ecology. Elders interact independently with products and people in their network of social connections. (b) An imbalanced elder ecology. Shifts in the ecology may be caused by the inability to independently and successfully use products, resulting in a gap between elder and environment. Elders need to rely on others for assistance and begin to contract services for household help. (c) An elder ecology sustained by future robotic products. Robotic products support multifaceted product interactions and activities. The elder has the same sphere of influence and quality of life as others in the ecology.

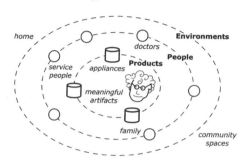

(a)

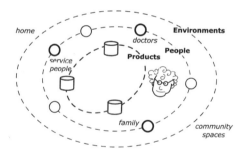

(b)

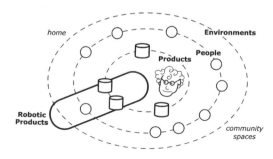

(c)

32

Unfortunately, restoring balance to the ecology is often not an easy proposition. For example, an elder could experience rapid decline as the result of illness or an accident. A physician might believe that a safe solution is to place the elder in an assisted living facility. However, the elder might not want to move there, the family might not have the financial means to do so, or no space may be available in an appropriate facility. Alternative measures are often put in place—and a more suitable solution may never be realized.

We propose that assistive robotic products can address the inevitable change and instability of the elder ecology, and provide balance while allowing the elder to retain independence and dignity. Figure 1c shows how future robotic products might reinstate the balance within the ecology by mediating among components that are not realizing the elders' values and goals.

## 2. METHOD

Ethnographic design research methods have been adapted from anthropology (Geertz, 1973), other social sciences (Csikszentmihalyi & Rochberg-Halton, 1981), and engineering (Button, 2000) to provide insights on how people form relations with products. Many of these studies have focused on how people create meaningful emotional relationships with products. Some have provided accounts of perceptions, attitudes, and beliefs about how products shape day-to-day activities. More focus on how people make sense of the formal and functional attributes of products is needed.

Design ethnographies focused on user–product interactions can contribute to the engineering and design of new products, artifacts, and services. Similar studies have provided ideas on how to bring technology to a traditionally technology-resistant user community (Salvador, Bell, & Anderson, 1999). These issues are particularly relevant when dealing with new products, including robotic products.

In this study, we examined elders' activities and interactions with products, an area comparatively uncharted in the literature. We conducted qualitative semistructured interviews and observations with 17 elders living in 15 private residences in the Pittsburgh and Chicago metropolitan areas. We investigated typical daily experiences for these participants, and focused on how products support or hinder activities for this population. To give more context to our findings, we talked with five experienced visiting nurses and social workers in a home healthcare program sponsored by a Pittsburgh hospital.

We identified two types of individuals: *well elders* who were mobile, cognitively intact, and able to maintain their households with relatively little help; and *declining elders*, who were experiencing either reduced mobility, cognitive impairment, or problems performing household maintenance tasks (Figure 2). Well elders ranged in age from 68 to 79. All 5 participants were

*Figure 2.* Distribution of participants.

| | Live in Family Home | Live in a Condominium | Live in an Elder-Specific Community | Live with Adult Children |
|---|---|---|---|---|
| Well elders | *n*= 2 | *n*= 2 | *n*= 1 | — |
| Declining elders | *n*= 4 | *n*= 2 | *n*= 5 | *n*= 1 |

women. In general, well elders spoke infrequently about the cognitive, physical, and emotional shortcomings associated with aging. Declining elders were older, ranging in age from 80 to 90. Nine declining elders were women, and 3 were men; one of the interviews was conducted with a man and wife, and one was conducted with a brother and sister. Declining elders spoke frequently about how aspects of aging changed their day-to-day activities, and how products they used to rely on were no longer usable or accessible.

## 2.1. Data Sources

We recruited participants by posting signs in surrounding neighborhoods and through networks of neighborhood volunteers. One of our first participants recommended additional participants in more advanced stages of decline. Because members of our research team have lived in the cities in which we conducted this study, we had several opportunities to see our participants beyond the context of the interview, observing them anecdotally in other places.

We conducted an in-depth interview with each of the 17 elders to gain an understanding of their experience of aging. We focused on elders' individual and collaborative routines, changes in activities and product use, and attitudes and perceptions that emerge as aging takes place. All of the interviews took place in elders' homes. They were accompanied by home tours where elders described their homes; showed us their favorite products, most useful products; and products that caused problems or were seen as needing improvement. We also observed key household events for periods of about 1 hr—primarily meal creation, meal consumption, and grocery and item storage. Occasionally, the participant was prompted to speak about what he or she was doing. Sessions were videotaped and photographed. We collected data over a 4-month period, and we generated field notes after each interview.

## 2.2. Analysis

Analysis of the data focused on creating participant profiles and reviewing and summarizing relations among participants, products, and the activities that specific products enabled or prevented. Products were coded using the Industrial Design Society of America (IDSA; 2002) standards for product categories. Activities were coded using the National Aging Information Center's (NAIC's) Activities of Daily Living and Independent Activities of Daily Living (NAIC, 1989), and the Extended Activities of Daily Living characterized by other research in the area of elder support (Mynatt et al., 2000; Figure 3).

## 3. FINDINGS

Our analysis of interviews with elders revealed the interconnected nature of their aging experience. Our ecological viewpoint proved to be useful in understanding a network of elders; the people they interact with; and their products, activities, and experiences all held in balance.

## 3.1. People

Our participants represent middle and upper middle class elders in America. They benefited from familial and financial support and had a variety of opportunities for social support and interaction through local senior centers and community groups. Our female participants also greatly outnumbered our male participants. Recent U.S. Census (2000) data revealed that women over age 65 compose 3% of the U.S. population, whereas men over age 65 account for only 2.6%. Most of the elders and all of the caregivers we spoke to were women. This may motivate somewhat of a rethinking of what values we need to support with appropriate products and services for this group.

Within our participant pool, we noted differences in perceived status, or how elders in a particular community defined their position relative to others. Elders who had status-affirming experiences such as college educations; travel; and lucrative, fulfilling careers told many stories about their lives when they were younger. They also spoke of current social events, volunteer work, and proudly providing familial assistance to aging parents in the past and adult children in the present. Their current activities were carefully chosen to project values that were important at the peak of their adult life.

This behavior is often heightened when an elder moves to a new community and is afforded the opportunity to create new social relationships. For example, Mrs. C. is a 69-year-old woman who had divorced and moved 10 years ago to a modern loft-type condominium not specifically for elders. Most of her previous adult life was spent living in large traditional country homes, some

*Figure 3.* **Product and activity codes used in ethnographic analysis.**

| Products | Activities |
|---|---|
| P1 Assistive products (hearing aids, walkers) | A1 Activities of daily living (bathing, dressing, eating, ambulation) |
| P2 Appliances and housewares (coffeemaker, etc.) | A2 Instrumental activities of daily living (meal preparation, household management, medicine management) |
| P3 Diagnostic equipment (blood sugar monitor) | A3 Extended activities of daily living (entertainment activities, social work, volunteer work) |
| P4 Entertainment products (stereo, television) | A4 Communication activities |
| P5 Medical equipment (medicine management) | |
| P6 Personal products and meaningful items | |
| P7 Services (cleaning, medicine management) | |
| P8 Technical products (computers, cell phones) | |
| P9 Transportation products (shuttle service, automobiles) | |

with more than 20 rooms. Describing this phase of her life, when she was working at a university and married to a successful businessman, she was proud of the quality of life she had and the ability to provide for others. She stated:

> It was like I ran a boarding house in town for those years. My children had visitors, our family frequently visited, and we had a steady stream of businessmen and visitors from the university. How did I find time to work?

Her move into a condominium environment was both a reduction in space as well as a lifestyle change. However, it was important for her to continue to reference the status and activities from her former life. Describing her home, she continued:

> It doesn't compare to any of my big old houses. But when my last child went off to college I thought, "What am I doing?" … and I thought … I'm just going to go someplace clean and modern. But all of my houses have been very old. So this is different for me. Most people have steel and leather furniture, and I just don't because it is not what I owned.

Referencing her furniture and possessions, she alluded to her favored country overstuffed chairs and the fact that her condominium is twice as large as others, affording her space and a place for her adult son (R), who only occasionally spent the night: " … and R thinks of it as home, his bedroom and bathroom."

Mrs. C. also had a baby grand piano, a product that dominated over one half of her loft-style living room, although she could no longer play it due to chronic back problems. Through her discussion of the piano, we understood it to signify her previous lifestyle, although it was no longer usable.

Elders with fewer status-affirming experiences were often less interested in using status to define who they were in current communities. They sometimes avoided community events and social outings. Others framed recollections of family assistance as tasks they were expected to take on, infringing on personal rights and freedom. For example, Mrs. G. is a 79-year-old woman who is relatively healthy and socially engaged. During her interview, she described her younger years:

> I didn't get an education, just high school. None of the girls in our family could get an education. We were all offered scholarships; we were all honors students and I couldn't get an education. My mother had three of us who were 18 months apart. The boys all got college educations. We went to work in the depression and supported them. I started to work at 14 and I got a good job during the war. I got a fantastic job with the government.

Signs of status were reinforced by how elders appeared during their interviews. Those of higher status showed and described clothing and jewelry collections, indicating that it was enjoyable to dress formally every day, wear makeup and jewelry, and have neatly coiffed hair. Interviews were generally conducted in the kitchen, and homemade food was offered. Those of lower status did not show clothing or the contents of closets during interviews. They wore loungewear or sweatsuits, occasionally had rumpled hair, and usually were not wearing jewelry.

We also found interesting patterns in our participants' patterns of product use that were indicative of activities and social interactions. For example, Mrs. A. had recently moved from upstate New York to live with her adult son in Pittsburgh. She purchased a cell phone to maintain ties with friends in her old neighborhood, but also to maintain her independence and identity within her son's home. Mrs. D. was mostly homebound and making do with her home appliances from nearly 50 years ago, but she had a new television and stereo. These had been selected and purchased for her by her children, in an attempt to improve her entertainment activities. Mrs. P. had a newly remodeled kitchen in her otherwise historic home. At first glance, it appeared to be modern and aesthetically pleasing, but we found that the location of the oven, dish-washer, and refrigerator had been carefully planned for someone who had trouble bending and reaching. Mrs. P.'s close friend, a young architect, had helped her with the plan so that she could easily continue entertaining and so-cializing—activities she valued. These examples indicate that buying and us-ing new products can be indicative of elders' values and the current state of their well-being in general.

These examples describe the elders we spent time with. They depict an ag-ing culture that is particular to urban residential neighborhoods. Many of our participants lived in close proximity in apartment and condominium com-plexes that if not created specifically for elders, housed mainly senior citizens. Many of them knew each other because their children attended school to-gether, or they were neighbors in former neighborhoods. We had the opportu-nity not only to observe them but to begin to understand the interconnectedness of their relationships in the elder community.

## 3.2. Products

How elders interact with products—whether they take the form of artifacts, services, or environments—plays a key role in defining the experience of ag-ing. As elders begin to decline, why they want products, how they use prod-ucts, and what they value about products changes. Elders are unique in their relationships with products for several reasons. First, elders generally have fewer reasons to make relations with new products as they age. Reduction in income, contraction in physical space, and reduction of social interaction limit opportunities to define relations with new products. Reduced or limited mo-bility also creates fewer opportunities for elders to interact with new products. Second, elders may adopt or ignore products based on how they reinforce per-sonal identity and values, particularly during the transition to smaller homes and new communities. For example, housewares, art objects, furniture, clothes, and jewelry provide a clear message to the community about who an elder is and even the status enjoyed in adult life. Third, sometimes products designed specifically for elders (particularly assistive products) are stigmatiz-ing and demeaning. These products are often not used at all or are modified to serve marginal uses. Product breakdowns like these create a gap between el-der and environment, sometimes resulting in danger, isolation, and eventually institutionalization.

### Why Do Elders Want Products?

We found that elders generally want products that match their aesthetic de-sires; that they use products that support their functional needs and abandon products that do not; and that the most important products are the ones that support elders' values of personal identity, dignity, and independence. After over a half century of interacting with products, many participants had adopted discriminating tastes. In our interviews, elders spoke at length about aesthetic qualities and personal meaning of cherished products. Products are traditionally used to define one's identity (possibly, in defeat of ageism), or to re-establish or maintain one's identity after relocating to a new home. For example, Mrs. A. is an 82-year-old artist who recently moved from her home in another state to live with one of her grown sons. The move forced Mrs. A. to reduce her possessions to those that were most important to her (her paintings and painting supplies) to fit in a home filled with her son's family's possessions. When comparing her current home to her previous home, Mrs. A. stated:

> The transition here has been very hard. Breezy was my *home*. Here, I *live* here. I used to cook a great deal. I did my own laundry. Now, everything is different. It's hard. Although my days are active, I stopped driving when I came here. That takes your everything, your independence away. It's, well, as I say, hard.

Mrs. A.'s home tour focused on her artwork in every room, rather than any of the family photos or possessions on display. In our time spent in the dining room, where the family ate together every night, she talked only about her artwork:

> That's one of mine. I would say that is something I did down at the beach. This is a view I would get from my house. I call it "Across the Bay." Why do you think I do short, wide paintings (laughs)? There's "Coney Island." There are a couple more paintings down here.

When she arrived at a shelf full of family photos and art objects, she chose only to describe a plaque that represented an art award that she had won. Through the objects that she chose to discuss during the interview, we felt that Mrs. A. was asserting her identity within her son's home.

Mrs. T. is 81-year-old woman who has lived in an elder high rise for 7 years. In the last 2 years, she had begun to decline rapidly, causing her son to become more concerned and to increase the amount of support and interaction he provided. During her interview, Mrs. T. spoke at length about an air conditioning unit purchased by her son:

> My son came in from Arizona, and he said "Mother, how could you live in here? It's so hot!" and he went to Home Depot and he bought it and he put in himself. I don't like the looks of the window, you know,

[referencing the connection to the window done in a crude manner with a large plastic hose] but … it *is* pretty [referencing the unit]. He paid over six hundred dollars for it … and then he needed another part, so he went out, back to Home Depot and bought another part. And you take it out in the wintertime. The janitor and the maintenance man will take it out in the wintertime when it gets cold. I'll have them put it back in and maybe they will do it right. But my son was in a hurry and he wanted to make sure that he got it for me.

Rather than describing its function, Mrs. T. chose to discuss the fine quality of the air conditioner, even describing it as pretty. We interpreted this exchange as being indicative of her pleasure in having her son contribute to a comfortable living environment.

### How Do Elders Use Products?

Elders use products because the functional aspects of products meet their current needs. Products are instrumental in completing a variety of daily activities. This reason differs from that of the young population, who often uses products because of stylistic considerations regardless of functionality. The elders in our interviews told many stories of how household appliances; transportation products; and communication products such as telephones, cell phones, and computers enabled them to help themselves, provide for family members and friends, and stay in touch with people in their social network.

Mrs. N. is an 80-year-old woman actively engaged in her community. She expressed pride in being able to help an acquaintance (P) in need, despite her own recent recovery from a bout of pneumonia:

Even though I am on hiatus [from many of her usual activities, due to pneumonia], once a week, I take P shopping. She is a person in the building. That's my helping work. Every Friday we go have our hair done, then we have lunch, then we do the shopping.

Whereas well elders mentioned product successes more frequently than product failures, declining elders talked at length about how the functional aspects of products and environments no longer served them. Eight of the 12 declining elders that we interviewed discussed how they could no longer easily make use of bathroom tubs, toilets, and fixtures; kitchen appliances, tables, and counters; telephones; clock radios; grocery carts; automobiles; and public transportation to support their basic needs.

Mrs. L. is a 79-year-old woman who suffers from depression, insomnia, neural degeneration, gastric reflux, and balance problems. She lived in a high-rise condominium for elders, but the design of her bathroom made it so inaccessible that she had great difficulty using it (Figure 4a). This was especially evident as she described the bathtub, shower, and hot and cold water faucets relative to the shortcomings of her own body. In describing the process of taking a shower, Mrs. L. commented:

> This apartment was made for old people, and they knew it when they made it. Yet why would they put that up so high? [referencing the height of the shower rod] I can just barely hang anything over there … it really is much too high. And another thing, I'm not so smart my dear, this faucet, I mean, you have to be a rocket scientist to use this faucet! I think it's very hard to use. Until I get it running right, I am ready to give up on it.

Unfortunately, environmental shortcomings such as the ones Mrs. L. described in the bathroom are enough to force elders to reduce their standards of personal hygiene.

Six of our participants showed and described modifications they had made to communications products and housewares, to increase accessibility. Our male participants described the process of modifying a product as an exciting challenge, whereas our female participants described it in terms of pure need. Modification results in a personalized product that is satisfying to use, as a conversation with Mr. G. about his personalized desk illustrates (Figure 4b):

> Yes, yes, I fixed that thing [the desk] up for myself. I did that for myself. It holds my envelopes, papers, pencils … everything is in there. I work on it periodically. If I see a box that looks better, I might take one down and put a new one up there instead.

As elders' bodies continue to decline, problems with products continue to be magnified; are less likely to be corrected; and ultimately result in messy, unsafe environments with more than one product to serve the same function. For example, Mrs. V. is an 81-year-old woman struggling with basic activities of daily living. During her interview, we noticed two radios on her bedside table, and we asked why (Figure 5a):

> You know what, [pointing to digital clock] I never used that as an alarm. I don't know how to set it. I use this one [pointing to analog clock]—but it is not any good … I have to get a new one. I use this one [pointing to digital clock] to look at the time.

*Figure 4.* (a) Mrs. L.'s water control in her shower was hard to understand and use and re-sulted in her bathing less frequently. (b) Mr. G. took great satisfaction in modifying his desk.

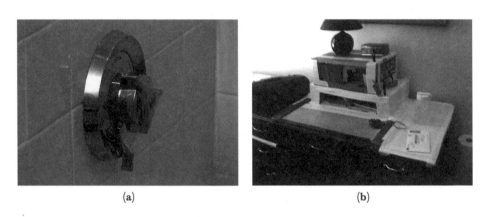

(a)    (b)

*Figure 5.* (a) A clock radio no longer serves someone with vision and muscular limitations, resulting in use of more than one product. (b) Bending and stretching to reach storage ar-eas becomes difficult, resulting in using the counter for storage and disarray in the kitchen.

(a)    (b)

The shortcomings of kitchen environments can make normal meal preparation difficult and even dangerous for elders. Mrs. G. kept busy baking a quiche and preparing fruit salad while we interviewed her. At one point, she needed to retrieve a container from a cupboard over the stove that was out of her reach. To do so, she took a broom and repeatedly jabbed it into the open cabinet until the item that she wanted fell out and on the floor. During this process, several other items fell out of the cabinet and landed on the stove, presenting a fire hazard. Mrs. G. also had trouble reaching items in the refrigerator, which was filled with containers of food in precarious locations and in various stages of deterioration.

Mrs. R. is a 90-year-old woman who is losing control over maintaining her home of 43 years. Her kitchen was also in a dangerous state of disarray (Figure 5b), which she repeatedly blamed on her laziness:

> My kitchen isn't fit to be seen … . [Mrs. R. starts cleaning passively, and interviewer tells her it is unnecessary] Well, I'll get around to it. It's not bad looking when you take all the stuff away. I'm just too lazy to do stuff … It's not bad when you can see all these plates, if I take time to clean. But I'm just maybe lazy or … (she trails off). [Interviewer asks if Mrs. R. uses the cupboards at all for storage anymore] Oh yeah, I've got my dishes. [Opens cupboard to reveal dishes and glassware wrapped in newsprint and plastic bags] And I use all this stuff in here. Well, if you look in here it's a mess. My son is after me, he says, "Do you want all that stuff on the floor? Put it in the basement or lift it up."

If elders can understand how assistive products can help them remain independent, they are likely to consider and accept using products such as hearing aids, dentures, canes, walkers, and wheelchairs. Without this understanding, there is resistance in acquiring and using assistive devices. We heard misconceptions about what assistive products are appropriate, and how they might be acquired. Out of the 12 declining elders that we interviewed, almost three fourths were not able to recognize the need for assistive products, and nearly one half had severe misconceptions about their purchase and use.

For example, through a conversation with Mrs. G., we interpreted that she was in denial about her failing health and ambulatory abilities. Although her doctor recommended surgery, she chose to avoid it for as long as possible:

> I said, "Well I'm not doing it!" [referencing her doctor's request that she have knee surgery]. I'm going to fight it—I'm going to work it out. I do it myself, I found out that if I use the topical medications … I can do it with exercising. I'm doing fine. Anyway, the doctor said I have to

have it done [the surgery] so I said that I'm up in years now, would it make any difference when I have it done? I'm going to be eighty, so therefore is it bad? He says it doesn't make any difference. Now I'm really actually not going to have it done unless I have to. I'm going to wait until I can't walk, if I can pull it off, and I don't know if I can pull it off. That's the unknown.

Mrs. V. had accepted her decline, but seemed wary of assistive devices and unwilling to think about making changes to her home. An 80-year-old declining elder, she had suffered cancer and serious infections related to surgery a few years ago and had used a walker during her long convalescence. When asked about using her walker, she commented:

Well, I must have used it that whole month. I couldn't walk [while recovering at home from cancer], and I did not want a wheelchair. The reason I did not want a wheelchair—I would become an invalid! It's so easy to become an invalid. See you don't realize it when you are young … .

This exchange suggests that although Mrs. V. was aware of her current need for assistance, she feared that by responding to it she would only decline more quickly. During her interview, we observed Mrs. V. having trouble with many of the products and environments within her home. She had ceased using many products altogether. Although simple modifications could have been made to drastically improve her quality of life (e.g., by asking her children or a cleaning service to help remove clutter and unused products), she seemed to be unaware of the benefits. When asked what changes she expected to make in her home in the next 5 years, she responded that if her husband would let her, she would like to build a patio off of the kitchen. Her inability to perceive the need for change not only presented an immediate physical hazard but also increased the likelihood that Mrs. V. would need to leave home for a professional care community.

Elders who accept physical and cognitive decline seem to be more willing to explore the use of appropriate assistive products. However, they often lack appropriate information to make decisions about what products will be the right ones. Mr. and Mrs. H. were aware of the shortcomings of old age and had begun to modify their living space by having custom cabinets built; placing an amplifier on the phone; and using specialized tools, such as electric can openers, in the kitchen. However, they told an interesting story about grab bars installed by the previous resident of their apartment:

When we moved into this place, these rails, well she [Mrs. H.] said, "they have got to go." It wasn't too long before we realized they are really useful, particularly for getting up from the john and the tub. [Interviewer asks why they wanted to remove the grab bars] Well, we didn't think we needed them. We were a young couple ten years ago! We were only in our 70s. Who needed them? The old lady who used to live here, her doctor son had them installed. It wasn't very long before we realized, it was a blessing to have them. Well [Mrs. H.] is pretty husky, and it is difficult, and she has trouble breathing. She has to use all her energy for breathing.

Despite the fact that Mr. and Mrs. H. had made many modifications to their home, it was only through direct experience of the grab bars that they realized their utility.

As elders continue to decline, they must begin to rely on family, friends, neighbors, or acquaintances to perform basic household tasks. Direct experiences with assistive products, such as the one described earlier, may be useful in illustrating the utility of assistive products and services. Fear and trepidation often accompany making the choice to try a new product or service, partly because of the fear of the unknown and partly because accepting these products and services is often seen as stigmatizing or as a sign of admitting defeat.

We have seen that elders choose products that please them aesthetically, products that will support them functionally, and products that are indicative of personal identity. In the next section, we explain how products support values important to those we interviewed.

### How Do Products Support Values for the Elder Population?

Our explorations of elder experiences revealed that independence and dignity were unanimously important to this population (see Dorfman, 1994). These values have behavioral and emotional aspects. Behavioral values are acted out in interactions with products and self-held standards for conduct and appearance—for example, being nicely dressed when interviewers arrived and offering home-cooked food as if we were guests. Emotional values are surrounded by intense feelings, and are often acted out in defensive arguments about particular behavior. Independence and dignity were evident in elders' stories about both products and activities. For example, Mrs. L. insisted on driving to do errands, although it was unsafe, rather than relying on her daughter, with whom she had a distant relationship. We interpreted this behavior as her way of asserting her independence from her daughter.

*Independence*, the state of being competent and self-supporting, is a common value for many adults, regardless of age or lifestage. For example, consider the independence cherished by a 16-year-old girl who has just learned how to drive and is experiencing the first of many interactions with a vehicle. For elders, a shifting of capabilities causes a particular reprioritization of products and activities that help them assert their independence. For example, many of our participants used cell phones to maintain social connections even as they had to rely on others to drive them to social functions. Independence was manifested behaviorally through product choices like these, and in actions like choosing to drive or to stay in a large home. Independence was manifested emotionally in the stories we elicited about how elders envisioned their future lives. A common response often began with, "My children have offered to help, but my hope is not to burden them."

*Dignity*, the state of being worthy of respect, is a particularly important value for elders. In our interviews, dignity was behaviorally manifested in an elder's desire to maintain a particular personal standard within the home or the community. For example, 9 of our participants had hired cleaning services to assist with household management. At least 3 of these elders had forged close friendships with the women who cleaned their homes. We interpreted these friendships as being indicative of managing the admission that help is needed at home with dignity.

Mrs. K. had a *Hemlock Society* publication hidden among her pile of magazines. Explicitly removing it from the stack and revealing it to us, she explained:

> I belong to this [the *Hemlock Society*]. Instead of being left to die in agony, I would rather go when I am still … able. I did not tell my family. My son would have a fit if he knew.

Rather than relying on her family to decide what to do when she experienced significant decline, Mrs. K. instead preferred freedom in making choices about the end of her life. We interpreted this as her way of asserting both her independence and her dignity over making final choices.

The exploration of these values and how they are manifested through interactions with products reveals how our participants prioritized products and activities that supported independence and dignity. Personal standards for products changed when the elder ecology shifted.

## 3.3. Environments

Environmental reduction is a critical component of the experience of aging and plays a role in the elder ecology. The *environmental press* theory has shown that reduction in environmental and social opportunities plays a significant role in the aging experience (Lawton, 1982). The home becomes especially important as time spent in travel, work settings, and other spaces declines. As elders move to smaller homes, they seek desirable surroundings in new and smaller spaces (Ward, La Gory, & Sherman, 1988). We observed three types of home environments. The first type was a participant's original home, inhabited for more than 15 or 20 years and occupied when the participant was younger and not subject to the shortcomings of old age. These were often expansive homes, filled with a collection of products and artifacts, showing few signs of change or modification. Often, several rooms within these homes were no longer used, or were used for the storage of random items. The second type was a small home, condominium, or apartment not specifically in an elder high rise. These spaces often showed signs of contraction and were inhabited as the result of a significant life event, such as divorce or the death of a spouse or child. The third type was an apartment or condominium in an elder community. These spaces were generally designed to support physical decline. Many of them had been further modified by our participants. Elder communities often had public laundry facilities and service staff, such as superintendents, to help residents. They also had community spaces where formal and informal social gatherings took place.

Many of the environments we saw in our interviews (even those designed and constructed specifically with elders in mind) did not fully accommodate participants' needs. Bathrooms and kitchens had particular shortcomings that impeded activities of daily living. Water controls are notoriously poorly designed (Norman, 1990). The problem was exacerbated for our population, as witnessed in our interview with Mrs. L. Her solution was to take fewer showers.

Kitchen environments often fail ergonomically. As elders decline, they may have limited physical mobility, which makes reaching kitchen surfaces, storage areas, and products situated within the kitchen quite difficult. In several kitchens (such as Mrs. R.'s discussed earlier), we found collections of foods, appliances, and other kitchen products in disarray on the kitchen counters.

In general, storage was a problem for most of our participants. Many could not see, let alone reach, upper and lower shelves of kitchen cabinets and storage closets. Only 3 participants had been proactive in reconciling this problem. One couple had custom cabinets built, and another participant had her cupboards rehung 8 in. lower so she could reach the shelves. Mrs. L.'s son built new shelves in her hall closet, creating an accessible space that she could use for customizable storage of toiletries and medications.

Usability and accessibility of the kitchen can make the act of preparing and eating food unsafe, tedious, and no longer enjoyable. We witnessed Mrs. L. making lunch in her kitchen. Her degenerative muscle disease made it very difficult for her to stand at the counter and use a paring knife to make a sandwich. Mrs. L. also had trouble getting in and out of a chair, so she had to continue to stand uncomfortably at the counter to eat lunch.

We used participants' comments about environmental shortcomings in bathrooms and kitchens to prompt a discussion of their perceived need for changes to home environments in the next 5 years. Not surprisingly, our participants were hesitant to describe changes in the foreseeable future. The majority of those we interviewed were clearly in the process of understanding and accepting various stages of decline. Most reported little need to make changes and instead described changes to the home in terms of aesthetic remodeling.

## 3.4. Activities and Experiences

We found that elders' interactions with products facilitated activities, mediated social interactions, and evoked experiences that contributed to their sense of self. Properly designed products and assistive products played a key role in helping elders undertake activities. Activities that promoted social interaction were extremely important for this group. They played a critical role in helping elders to remain healthy, happy, and independent. Our participants cited a large range of activities that helped them stay engaged socially. The inability to participate in these activities resulted in contraction of their social space.

Our participants undertook a variety of activities, including family outings, visits to friends' homes, meals, volunteer activities, and religious and community events. Many activities mentioned were not explicitly described as social, but were implicitly social in nature. These included lifelong learning classes, exercise classes, doctor visits, and assistance to neighbors in the community.

We believe that particular living arrangements supported frequent social interactions with both family members and the community. Mrs. G. said she "maintained" two households—one with her estranged husband and another with her daughter and granddaughter. Mrs. G. spent most of her time at her daughter's house, providing "assistance" in buying food and preparing meals. We observed her working in her kitchen during our visit. The kitchen cupboards and the refrigerator were in a state of general disarray. Several times while cooking, Mrs. G. neglected to clean the utensils before placing them back in the drawers. These observations suggest that the relationship was more a social than a practical necessity for Mrs. G.'s daughter and granddaughter. Mrs. G. may have been acting beyond her capabilities and possibly straining the very relationship she believed to be helping.

Volunteering and helping others are activities that strongly define an elder's sense of self-identity. For example, Mrs. C. participated in four different volunteer activities. She was a founding member of a cooperatively managed used bookstore, a church trustee, a trustee at a credit foundation, and a board member for a local school organization. In addition to participating in these activities, she helped others to participate by driving to and from events.

We saw that decline, mediated by breakdowns in product use, drastically reduced elders' activities. At that point, many activities not ostensibly intended for social interaction in middle age became valued points of engagement in old age. These included activities such as doing laundry in a communal facility, receiving a visit from a home nurse, or participating in exercise and physical therapy classes. Nine of the 17 elders that we interviewed participated in at least one such activity every week, if not every day. These activities were described in social rather than functional terms. They often provided an opportunity to leave the house, meet peers, and make light of aches and pains. Many physical therapy and exercise classes took place at senior community centers rather than hospitals, further emphasizing their social nature. Often, exercise made our participants feel young and desirable. Mr. H., who exercised nine times a week, proudly professed, "I'm a jock, and I get to spend lots of time with widows!"

Gradual, yet substantial, decline in abilities can have especially damaging effects on social interaction, because elders can simply give up. Figure 6 compares the number of times activities of daily living, instrumental activities of daily living, extended activities of daily living, and communication activities were mentioned during interviews. Declining elders mentioned basic and instrumental daily activities more frequently, consistently describing disappointment in no longer being able to successfully undertake a given activity.

Mrs. L. was a poignant example of this disappointment. Her physical decline was recent, but rapid and extensive. At the beginning of the interview, Mrs. L. commented:

Everything has changed. I mean, my life is completely different [since the onset of multiple conditions]. But I still try to go and do. My neighbor has asked me to go to lunch. I see her, and she says, "When are we going?" I hate to have to tell her that it is just too hard.

Mrs. L.'s situation is interesting to interpret. Over the course of the interview we began to understand how difficult social interaction had become for her. She occasionally drove to the grocery store and the beauty salon, went on outings with her family, and maintained relationships with a few women in her building. However, she spent most of her day watching TV, despite the fact that her friends made repeated efforts to engage her socially. Near the end of the interview, she commented that socializing was becoming too much of an effort. We felt this indicated that Mrs. L. still desired social interaction but that the shifts in her ecology were making social engagements harder and harder for her to undertake. These kinds of situations can lead to isolation or even danger. Mrs. L. could eventually give up attempting social engagements, although they may be quite feasible with proper assistance.

*Figure 6.* Frequency that well and declining elders mentioned different activities.

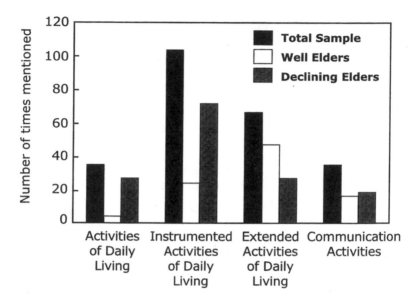

Although assistive products are often described as facilitating the activities of daily living by making those activities accessible to people with physical or cognitive disabilities (Fernie, 1991), our interviews showed that assistive products could also play a powerful role in helping elders to create socially engaging experiences. Mrs. T., an 82-year-old woman suffering from degenerative muscular disease, provided a clear example of how an assistive product can serve social needs. Mrs. T. could barely swallow, walk, or get in and out of a chair, and she fell several times a day. Despite of all this, she insisted on taking part in the informal social gathering each afternoon in the courtyard of her building. In the past year, it was only after acquiring a walker that she could even attempt this activity. Although she relied on others for almost all aspects of personal and household management, she enjoyed having the agency to partake in this social activity. We believed that her walker was valued not in terms of facilitating her mobility but instead in terms of creating opportunities for her to socially engage with the world.

## 3.5. Summary of Findings

Within the framework of the ecology of aging, we found that for elders, products can meet (or be adapted to) newfound needs, but especially when elders decline, they can serve as a source of frustration and hardship. For example, a walker represents the chance to maintain one's autonomy, but if it is heavy, ugly, and cannot be used outside, its acquisition can instead prompt greater reliance on others or, if getting help is impossible or undesirable, isolation. This example reflects our major finding that many products used in and around the home can represent values for elders that they do not represent for those at younger ages and that products can have some unintended consequences for elders as well. For example, the cell phone begins to substitute for the car as a way to contact others. We believe our results point to considerations for ecologically sensitive design of robotic assistive products.

Our study has several limitations such that further observations are warranted. In particular, our study used a small sample. Also, it was geographically bounded by two Midwestern cities and physically bounded by our participants' homes. Increasing the diversity of participants could improve our understanding of how to design products for a broad population. For instance, one of our participants, a retired engineer who designed his own sleep apnea device, will most likely react differently to a robotic product than a female participant who never learned how to drive. Robotic products may need to have familiar appliance-type product forms that are generally acceptable to a variety of users, or robotic technology may be added to familiar and widely used products and environments. Increasing sample size and diversity will also increase the diversity of observed environments. Each home has idiosyncrasies that differentiate it from other homes as well as from institutionalized settings (Michelson & Tepperman, 2003). Finally, longitudinal studies of the transition from homes to institutions could provide more clues about how elders' lives unfold.

## 4. ROBOTIC PRODUCTS TO SUPPORT THE ECOLOGY OF AGING

Robotic technologies have now reached a level of sophistication that makes deployment in the home feasible. Conceiving of the aging experience as a set of interrelated independencies within an ecological system is a critical first step for designing robotic products suitable for elders. Our ecological framework exposes the many important contingencies that exist in designing products to provide a good quality of life for elders and those who provide care for them.

Later, we use our data to suggest guidelines for the design of appropriate robotic products for elders in their homes. Our guidelines are necessarily broad and preliminary. Technology development is rapid. Our description of the experience of aging is not frozen in time. The experience of aging, and technologies, will certainly change in the next decade. Although we can describe the changes and dependencies we witnessed in the ecology of today's elders, we cannot confidently project those needs 5 to 10 years into the future. We believe, however, that our design guidelines can serve as initial "rules of thumb" for addressing design goals for robotic product development for elders in the home.

In the future, robotic product design might become as common as consumer product design. As we learn more about this new area of product development, we will generate more specific principles and begin to populate the landscape with detailed case studies of robotic products that chart the course for future research and design.

Our research, and that of those in the technical research community (Kawamura, Pack, Bishay, & Iskarous, 1996; Schraft, Schaeffer, & May, 1998), has begun to identify specific needs and values future robotic technologies might support for this population. Some of these include support in managing the home; maintaining personal and household supply; monitoring and providing ambulatory support; and most important, providing communication and social interaction. Our ethnography revealed needs ranging from making meals to having mail read. Although these are good starting points for generating product ideas, we also advocate taking a holistic approach and thinking about how future robotic technology might impact the elder ecology as a whole. A robotic product has the potential to be a powerful technology—engaging with it will have different, unexpected, and potentially negative results. The following guidelines begin to chart a course for successfully integrating new products into the lives of elders.

## 4.1. Design Guidelines

### Robotic Products Must Fit the Ecology as Part of a System

Robotic products designed for an elder ecology must be conceived of as part of a larger system of existing products and environments that serve elders and others in the ecology. When designing, first-order effects (such as fit) and second-order effects (such as social and cultural implications) must be considered. Familiar product forms with augmented product functionality will fit the system and maximize early product adoption.

As part of a larger system, future robotic products need to offer diverse interactions to many different groups of people. One of the most formidable tasks for designers is to conceive of and construct the interactions that will take place between elders, future robotic products, and others in the elder ecology. We will not be able to predict how the social situations and surrounding physical context will respond to, and create unanticipated uses of, future robotic products within an elder ecology. For example, an elder might use a robotic product to communicate with her daughter, primary physician, physical therapist, and even pharmacist. Rather than a product that is used by one person, such a product offers opportunities for interaction among all members of the elder ecology. These products should maximize fit by allowing the elder to experience the same power, control, and agency as others in the elder ecology.

Part of the designer's task is to understand how these products situate within the ecology and what kinds of second-order effects will occur. For example, refrigerators were designed to keep things cool, but have also informally become a vertical communication space through the use of refrigerator magnets for notes, artwork, cartoons, and family communications. Similarly, we observed a robot designed to deliver medicines in a hospital. The robot was repeatedly decorated with flowers and stickers by hospital patients as it made its rounds. In thinking about product function and aesthetics, it will be important to consider what intended and unintended opportunities for communication, sharing, and new experiences future robotic products will offer.

## Robotic Products Must Support the Changing Values of Those Within the Ecology

Robotic products designed for an elder ecology should uphold the important values of independence and dignity. They should also allow for different prioritization of these values. When designing product functionality, allow for basic needs as well as higher needs to be addressed. Combining appropriate and accessible functionality along with aesthetic considerations will support values and sustain product use.

Our description of the experience of aging has shown that change is inevitable for elders—change in their physical and mental abilities, and change in the local environment in which they live. These changes account for endless shifts in values, individual prioritizations of values, and endless differences in the elders we interviewed. Robotic products need to support these shifts in values by responding to different decisions and actions in any given situation.

Interactions with these products should not detract from the elder's independence and dignity. They should allow elders to do as much as possible, by enabling people, instead of the system, to initiate most of the interactions with the product. To help the elder feel comfortable using new technologies, future robotic products should allow for flexible and accessible use. All interactions should be designed to support use and maintain the values of the widest variety of people. The Universal Design Principles (2002) developed in the last decade have ensured that product functionality serves the widest group of human needs. It is essential that products also serve the widest group of human values by supporting the many forms of independence and agency within the ecology.

Future robotic products should also support independence and dignity by maintaining the best quality of life that the elder has known, or perhaps, improving it significantly. In evaluating future robotic products and their effect on the elders who use them, we will need to move beyond cognitive and ergonomic issues and deeply consider the human–machine relation, focusing on the network of relations, values, and motivations involved in an ecological view of adopting new robotic products.

### Robotic Products Must Be Functionally Adaptive

Robotic products designed for an elder ecology must be adaptive. They should provide a solution for today as well as solutions that will support future change. When designing products, allow for extensible and mutable forms and functions.

Robotic products that are situated successfully within an ecology will support change and growth, fostering new activities for everyone to be involved in, as well as conditions for adaptation; new knowledge generation; and creative problem solving rather than accepting less than desirable situations. One pertinent example from our earlier research in a nursing home can provide some insight. The community received two desktop computers with Internet access, and placed them in a public room on the first floor of the building. Although the staff at first believed that there would be little uptake of this new technology, they quickly learned that just the opposite was true. A few knowledgeable staff and residents were quickly elevated to the level of "local experts," and signup sheets had to be created to regulate fair use of the equipment. Residents asked for lessons on how to use the computer, and eventually, a technical staff member was hired. One year later, the presence of the computers had facilitated new social interactions. Residents had taken over the production of the community newsletter, and other project plans were underway. As a result of the new technology, all of the components of the ecology were evolving. In addition, the technology was evolved by the technical staff member to make sure it continued to serve the ecology.

These design guidelines can be generally applied to the design of assistive robotic products. Specific design solutions will bring more knowledge to this new area of product development. Figure 7, for example, shows how our generative findings can lead to more specific design recommendations. As more specific principles are discovered, more detailed case studies will follow.

## 4.2. An Ecological Approach to the Design Process

Design is the human activity of conceiving, planning, and making the artifacts, services, and environments that contribute to the best quality of human

*Figure 7.* Summary of preliminary design guidelines and example design recommendations.

| Design Guideline | Example Design Recommendation |
|---|---|
| Robotic products must fit the ecology as part of a system. | Consider scale and footprint |
| | Consider placement in the home environment |
| | Make the product portable and usable beyond home context |
| | Use familiar product forms to inspire early adoption |
| Robotic products must support the migrating values of elders and others within the ecology. | Provide a natural, "walk up and use" interface |
| | Allow the user to initiate product interactions |
| | Provide more than one choice to complete any given task |
| | Provide options for aesthetic appearance |
| Robotic products must be functionally adaptive. | Provide multimodal input and consistent lightweight output |
| | Support universal access and use by the largest number of people in the ecology |
| | Provide mutable functionality for different users and contexts |

life. The activity of designing is common to a number of disciplines: engineering, science, art, rhetoric, and others. However, there are fundamental differences between how designers approach design (searching for solutions to witnessed problems) and how those in the harder sciences approach design (engineering new technologies in usable forms). The ecological framework presented here can provide a checks-and-balances system for all of the disciplines involved in the act of designing.

When designing new robotic products, we will need to make judgments about what technologies to pursue, what systems to make, and how to consider context when designing artifacts and services. Design problems of this kind are characterized as "wicked problems"; a class of social system problems that are ill formulated, confusing, and influenced by many decision makers. Wicked problems are indeterminate, meaning that there are no definitive solutions and more than one appropriate solution to any given problem (Buchanan, 1995).

Our ecology of aging brings a human-centered focus to designing future robotic products for the aging population. It bridges the gap between the many disciplines involved in the act of design. It allows for a number of solutions to address the needs and values associated with aging in place, by providing an approach for indeterminate and multidimensional problems. The ecology of aging allows us to formulate and test solutions while continually deepening our definition of the design problem. It allows us to take a checks-and-balances approach to design the best solution for all the components of the ecology—people, products, and activities taking place in particular environments. The ecology shows that technology is not the only influence in designing new products—social dynamics, economics, and environmental issues also play an important role.

## NOTES

*Acknowledgments.* We thank John Zimmerman, Michael Gillinov, and our anonymous reviewers for their helpful comments on the article. We thank Ann Kauth for her assistance with data collection.

*Support.* This research was supported by a grant from the National Science Foundation grant IIS-0121426.

*Authors' Present Addresses.* Jodi Forlizzi, Human-Computer Interaction Institute, Carnegie Mellon University, Pittsburgh, PA 15213. E-mail: forlizzi@cs.cmu.edu. Carl DiSalvo, School of Design, Carnegie Mellon University, Pittsburgh, PA 15213. E-mail: disalvo@andrew.cmu.edu. Francine Gemperle, Institute for Complex Engineered Systems, Carnegie Mellon University, Pittsburgh, PA 15213. E-mail: gemperle@cmu.edu.

*HCI Editorial Record.* First manuscript received December 2, 2002. Revision received June 2, 2003. Accepted by Sara Kiesler and Pamela Hinds. Final manuscript received September 16, 2003. — *Editor*

## REFERENCES

Bailey, M. (1986). *Golden years, tarnished hours: Ethnography of two elderly residences in the Midwest.* Unpublished doctoral dissertation, University of Michigan.

Bell, G. (2001). Looking across the Atlantic: Using ethnographic methods to make sense of Europe. *Intel Technology Journal, Q3,* 1–10.

Bell, G. (2002). *Making sense of museums* (Technical Report). Portland, OR: Intel Labs.

Buchanan, R. (1995). Wicked problems in design thinking. In R. Buchanan & V. Margolin (Eds.), *The idea of design* (pp. 3–20). Cambridge, MA: MIT Press.

Button, G. (2000). The ethnographic tradition and design. *Design Studies, 21*(4), 319–332.

Computing Research Association. (2003). *Grand research challenges in information systems.* Washington, DC: CRA Press.

Csikszentmihalyi, M., & Rochberg-Halton, M. (1981). *The meaning of things*. Boston, MA: Cambridge University Press.

Dorfman, K. A. (1994). *Aging into the 21st century: The exploration of aspirations and values*. New York: Brunner/Mazel Inc.

Dubowsky, S., Genot, F., Godding, H., Kozono, A., Skwerrsky, H., Yu, H., et al. (2000). PAMM—A robotic aid to the elderly for mobility assistance and monitoring. *Proceedings of the IEEE International Conference on Robotics and Automation*. New York: IEEE.

Fernie, G. (1991). Assistive devices, robotics, and quality of life in the frail elderly. In J. E. Birren, J. E. Lubben, J. C. Rowe, & D. E. Detuchman (Eds.), *The concept and measurement of quality of life in the frail elderly* (pp. 142–167). New York: Academic.

Geertz, C. (1973). *The interpretation of cultures*. Chicago, IL: Basic Books.

Golant, S. M. (1984). *A place to grow old: The meaning of environment in old age*. New York: Columbia University Press.

GuideCane. (2002). *The Guide Cane*. Retrieved December 1, 2002, from http://www-personal.engin.umich.edu/~johannb/GC_News/GC_News. html

GUIDO. (2002). *Guidosmart Walker*. Retrieved December 1, 2002, from http://www.haptica.com/id2.htm

Guralnick, J., LaCroix, A., Abbott, R., Berkman, L., Satterfield, S., Evans, D., et al. (1993). Maintaining mobility in late life. *American Journal of Epidemiology, 137,* 845–857.

Haptica. (2002). *The Haptica Walker*. Retrieved December 1, 2002, from http://www.haptica.com/whatwedo/walker.html

Harris, M. (1979). *Cultural materialism: The struggle for a science of culture*. New York: Vintage.

Hazan, H. (1994). *Old age: Constructions and deconstructions*. Cambridge, England: Cambridge University Press.

Hirsch, T., Forlizzi, J., Hyder, E., Goetz, J., Stroback, J., & Kurtz, C. (2000). The ELDeR Project: Social and emotional factors in the design of eldercare technologies. *Proceedings of the CUU 2000 Conference on Universal Usability*. New York: ACM.

Industrial Design Society of America. (2002). *IDSA home*. Retrieved December 1, 2002, from http://www.idsa.org.

Kawamura, K., Pack, R., Bishay, M., & Iskarous, T. (1996). Design philosophy for service robots. *Robotics and Autonomous Systems, 18,* 109–116.

Lankenau, A., & Röfer, T. (2001, March). A versatile and safe mobility assistant. *IEEE Robotics and Automation Magazine, 8,* 29–37.

Lawton, M. P. (1982). Competence, environmental press, and the adaptation of older people. In M. P. Lawton, P. G. Windley, & T. O. Byerts (Eds.), *Aging and the environment: Theoretical approaches* (pp. 5–16). New York: Springer.

Living at Home. (2002). *Interviews with staff, Living at Home Program*. Pittsburgh, PA: University of Pittsburgh Medical Center.

Mann, W. C., Hurren, D., Tomita, M., & Charvat, B. (1995). An analysis of problems with walkers encountered by elderly persons. *Physical & Occupational Therapy in Geriatrics, 13*(1/2), 1–23.

Marcoux, J.-S. (2001). The "Casser Maison" ritual: Constructing the self by emptying the home. *Journal of Material Culture, 6,* 213–235.

Margolin, V. (1997). Getting to know the user. *Design Studies, 18,* 227–235.

Mazo, M. (2001, March). An integral system for assisted mobility. *IEEE Robotics and Automation Magazine, 8,* 46–56.

McCormack, M., & Forlizzi, J. (2000). Listening to user experience: Integrating technology with proactive wellness management. *Proceedings of the PDC 2000 Participatory Design Conference.* Palo Alto, CA: CPSR.

Michelson, W., & Tepperman, M. (2003). Focus on home: What time-use data can tell us about caregiving to adults. *Journal of Social Issues, 59,* 591–610.

Morris, A., Donamukkala, R., Kapuria, A., Steinfeld, A., Talbot-Matthews, J., Dunbar-Jacob, J., et al. (2002). *A robotic walker that provides guidance* [White Paper]. Retrieved December 1, 2002, from http://www-2.cs.cmu.edu/~thrun/papers/thrun.robo-walker.html

MOVAID. (2002). *MOVAID.* Retrieved December 1, 2002, from http://www-arts.sssup.it/research/projects/MOVAID/default.htm

Mynatt, E. D., Essa, I., & Rogers, W. (2000). Increasing the opportunities for aging in place. *Proceedings of the CUU 2000 Conference on Universal Usability.* New York: ACM.

Nardi, B. A., & O'Day, V. L. (1999). *Information ecologies: Using technology with heart.* Cambridge, MA: MIT Press.

National Aging Information Center. (1989). *Limitations in activities of daily living among the elderly* [White Paper]. Retrieved December 1, 2002, from http://www.aoa.gov/aoa/stats/adllimits/httoc.htm

NavChair. (2002). *The NavChair: An assistive navigation system for wheelchairs based upon Mobile Robot Obstacle Avoidance.* Retrieved December 1, 2002, from http://www-personal.engin.umich.edu/~johannb/navchair.htm

Netting, R. (1986). *Cultural ecology* (2nd ed.). Prospect Heights, IL: Waveland.

Norman, D. A. (1990). *The design of everyday things.* New York: Basic Books.

Prassler, E., Scholz, J., & Fiorini, P. (2001, March). A robotic wheelchair for crowded public environments. *IEEE Robotics and Automation Magazine, 8,* 38–45.

RAID Workstation. (2002). *RAID Workstation.* Retrieved December 1, 2002, from http://www.oxim.co.uk/std.html#RAID

RAIL. (2002). *RAIL. Robotic Aid to Independent Living.* Retrieved December 1, 2002, from http://www.robotics.lu.se/Robotics/research/projects/RAIL/RAIL.html

Salvador, T., Bell, G., & Anderson, K. (1999). Design ethnography. *Design Management Journal, 10*(4), 35–41.

Scherer, M. J., & Galvin, J. C. (1994). Matching people with technology. *Rehabilitation Management, 7*(2), 128–130.

Schraft, R. D., Schaeffer, C., & May, T. (1998). Care-O-bot™: The concept of a system for assisting elderly or disabled persons in home environments. *Proceedings of the 24th IEEE IECON,* Vol. 4. Aechen, Germany: IEEE.

Silverman, P. (1987). Community settings. In P. Silverman (Ed.), *The elderly as modern pioneers* (pp. 234–262). Bloomington: Indiana University Press.

Universal Design Principles. (2002). *The Center for Universal Design-What is Universal Design?* Retrieved September 1, 2003, from http://www.desugb/ncsu.edu:8120/cud/univ_design/princ_overview.htm

U.S. Census. (2000). *Census Bureau Homepage.* Retrieved September 1, 2003, from http://www.census.gov

Ward, R. A., La Gory, M., & Sherman, S. R. (1988). *The environment for aging.* Tuscaloosa: University of Alabama Press.

Wheelesley. (2002). *Wheelesley: Development of a Robotic Wheelchair System.* Retrieved December 1, 2002, from http://www.ai.mit.edu/peole/holly/ wheelesley/

HUMAN-COMPUTER INTERACTION, 2004, Volume 19, pp. 61–84

# Interactive Robots as Social Partners and Peer Tutors for Children: A Field Trial

## Takayuki Kanda, Takayuki Hirano, and Daniel Eaton
*ATR Intelligent Robotics and Communication Laboratories*

## Hiroshi Ishiguro
*Osaka University*

**Takayuki Kanda** is a computer scientist with interests in intelligent robots and human-robot interaction; he is a researcher in the Intelligent Robotics and Communication Laboratories at ATR (Advanced Telecommunications Research Institute), Kyoto, Japan. **Takayuki Hirano** is a computer scientist with an interest in human–robot interaction; he is an intern researcher in the Intelligent Robotics and Communication Laboratories at ATR, Kyoto, Japan. **Daniel Eaton** is a computer scientist with an interest in human–robot interaction; he is an intern researcher in the Intelligent Robotics and Communication Laboratories at ATR, Kyoto, Japan. **Hiroshi Ishiguro** is a computer scientist with interests in computer vision and intelligent robots; he is Professor of Adaptive Machine Systems in the School of Engineering at Osaka University, Osaka, Japan, and a visiting group leader in the Intelligent Robotics and Communication Laboratories at ATR, Kyoto, Japan.

## CONTENTS

## ABSTRACT

Robots increasingly have the potential to interact with people in daily life. It is believed that, based on this ability, they will play an essential role in human society in the not-so-distant future. This article examined the proposition that robots could form relationships with children and that children might learn from robots as they learn from other children. In this article, this idea is studied in an 18-day field trial held at a Japanese elementary school. Two English-speaking "Robovie" robots interacted with first- and sixth-grade pupils at the perimeter of their respective classrooms. Using wireless identification tags and sensors, these robots identified and interacted with children who came near them. The robots gestured and spoke English with the children, using a vocabulary of about 300 sentences for speaking and 50 words for recognition. The children were given a brief picture–word matching English test at the start of the trial, after 1 week and after 2 weeks. Interactions were counted using the tags, and video and audio were recorded. In the majority of cases, a child's friends were present during the interactions.

Interaction with the robot was frequent in the 1st week, and then it fell off sharply by the 2nd week. Nonetheless, some children continued to interact with the robot. Interaction time during the 2nd week predicted improvements in English skill at the posttest, controlling for pretest scores. Further analyses indicate that the robots may have been more successful in establishing common ground and influence when the children already had some initial proficiency or interest in English. These results suggest that interactive robots should be designed to have something in common with their users, providing a social as well as technical challenge.

---

# 1. INTRODUCTION

## 1.1. Research on Partner Robots

The development of humanoid robots such as Honda's ASIMO (Hirai, Hirose, Haikawa, & Takenaka, 1998) and interactive robots such as Sony's AIBO® (Fujita, 2001) and Kismet (Breazeal & Scassellati, 1999) has spawned a new area of research known as interactive robotics. These are not robots performing simple iterative tasks in factories or using specific tools in professional services such as surgical or military tasks (Thrun, 2004). Rather, this new wave of research is exploring the potential for *partner robots* to interact with people in daily life. Our research explores some fundamental problems in this new field.

Several researchers and companies have endeavored to realize robots as partners for people, and the concept of a partner robot is rapidly emerging. Typically equipped with an anthropomorphic body and various sensors used to interact with people naturally, the partner robot acts as a peer in everyday life. A humanoid robot, for example, guides office visitors by speech and with a hand-gesture recognition mechanism (Sakagami et al., 2002). For the home environment, NEC Corporation (2002) developed a prototype of a personal robot that recognizes individuals' faces, entertains family members with its limited speech ability, and performs as an interface to television and e-mail. Partner robots have also appeared in therapeutic applications. For example, Dautenhahn and Werry (2002) are applying robots to autism therapy. As these examples show, partner robots are beginning to participate in human society by performing a variety of tasks and functions.

Eliza was the first computer agent that established a relationship as a partner (Weizenbaum, 1966). People tried to interact with Eliza without necessarily having a specific task or request in mind. They sometimes made brief small talk and at other times engaged deeply in conversation. As Reeves and Nass (1996) discovered, humans unconsciously behave toward such a computer as if it were human. In recent robotics research, several pioneering studies have suggested that humans also can establish relationships with pet robots. Many people actively interact with animal-like pet robots. For example, people have adapted to the limited interactive ability of the robot dog, AIBO (Friedman, Kahn, & Hagman, 2003; Fujita, 2001). Furthermore, pet robots have been used successfully in therapy for the elderly, with some positive effects of their usage confirmed in long-term trials (Wada, Shibata, Saito, & Tanie, 2002).

## 1.2. Social Relationships Over Time

Recognizing the other person's identity, discovering similarities, and finding common ground are key issues in cementing social relationships. As Isaacs and Clark (1987) proposed, when people first meet, they gradually establish common ground through conversation. Empirical studies have shown that interlocutors adapt their speech to each other's attitudes and experience, weighing each other's perspectives when listening and making themselves understood (Fussell & Krauss, 1992). In forming satisfying and stable intimate relationships, they may even find similarities in their partner that do not exist in reality and tend to assume that their partner is a mirror of themselves (Murray, Holmes, Bellavia, Griffin, & Dolderman, 2002).

This evidence shows the importance of finding common ground in establishing relationships. However, relationships among people evolve over time (Hinde, 1988), and we believe people's attitude toward technological artifacts and their relationship with them also evolves over time. Little previous research has focused on long-term relations between individuals and computer systems in general or partner robots in particular. Short-term and long-term analyses must be carried out to evaluate partner robots. With respect to short-term experiments, many evaluation methods and systems have been proposed within the field of human–computer interaction and robotics. For instance, Quek et al. (2002) developed a gesture recognition-based system to analyze multimodal discourse. In robotics, Nakata, Sato, and Mori (1998) analyzed the effects of expressing emotions and intention. We have also performed several similar experiments, such as examining the effects of behavior pattern on impressions (Kanda, Ishiguro, & Ishida, 2001; Kanda, Ishiguro, Ono, Imai, & Nakatsu, 2002). However, in short-term human–robot interaction, we can only observe first impressions and the initial process of establishing relationships.

Some previous research has stressed the importance of long-term studies. Fish, Kraut, Root, and Rice (1992) evaluated a videoconferencing system and analyzed the transition of system use during 1 month of experimentation. Petersen, Madsen, and Kjær (2002) reported on the process of gaining experience with a new television system. These studies showed that the relation between human and agent is likely to change over time, just as interhuman relationships do. Therefore, it is vital to observe relationships between individuals and partner robots in an environment where long-term interaction is possible. The result of immersing a robot in an environment that demands ongoing participation is likely to be entirely different from that of exhibiting the robot in a public place like a museum, where the people who interact with it are transient.

## 1.3. Technologies for Creating Human–Robot Relationships

As previous research on interpersonal communication indicates, it is vital that two parties recognize each other for their relationship to develop. We cannot imagine having human partners or peers who cannot identify us. It is because we are able to identify individuals that we can develop a unique relationship with each of them (Cowley & MacDorman, 1995; Hinde, 1988). Although person identification (ID) is an essential requirement for a partner robot, current visual and auditory sensing technologies cannot reliably support it. Therefore, an unfortunate consequence is that a robot may behave the same with everyone.

Given only visual and auditory sensors, it is difficult to implement a person ID mechanism in robots that works in complex social settings. Many people may be talking at once, lighting conditions may vary, and the shapes and colors of the objects in the environment may be too complex for current computer vision technologies to function. In addition, the method of ID must be robust. Misidentification can ruin a relationship. For example, a person may be hurt or offended if the robot were to call the person by somebody else's name. To make matters worse, partner robots that work in a public place need to be able to distinguish between hundreds of people and to identify nearby individuals simultaneously. For instance, consider a situation involving people and robots working together in an office building, school, or hospital.

Besides their ability to identify and recognize others, robots should have sufficient interaction ability. In particular, human interaction largely depends on language communication. Whereas speaking is not so difficult for the partner robot, listening and recognizing human utterances is one of the most difficult challenges in human–robot interaction. Although some of the computer interfaces successfully employ speech input via microphone, it is far more difficult for the robots to recognize human utterances, because the robots suffer from noise from surrounding humans (background talk) and the robot body (motor noise). Little research has reported the solutions to this serious problem. We cannot expect ideal language perception ability like humans. However, we believe that robots can maintain interaction with humans, if they can recognize other human behaviors, such as distance, touching actions, and visual movements, in addition to utterances.

People have bodies that afford sophisticated means of expression through diverse channels. We believe that a robot partner, ideally, would have a humanlike body. A robot with a human-like body allows people to intuitively understand its gestures, which in turn causes people to behave unconsciously as if they were communicating with a human. These effects have even been observed with screen-bound agents that move and point (Isbister, Nakanishi, Ishida, & Nass, 2000). We believe that this anthropomorphic basis not only supports the embodiment of computer interfaces (Cassell et al., 1999), but also enables their grounding in social relationships (Cowley & MacDorman, 1995). Eye contact, gesture observation, and imitation in human–robot interactions greatly increase people's understanding of utterances (Ono & Imai, 2000). Close synchronization of embodied communication also plays an important role in establishing a communicative relation between the speaker and listeners (Ono, Ishiguro, & Imai, 2001). We believe that in designing an interactive robot, its body should be based on the human body to produce the most effective communication.

When partner robots are involved in people's daily life, they will take on certain roles and contribute to humans based on their skills. Apparently, a robot that is skilled at a single or limited set of tasks cannot satisfy the designation of partner. For example, a museum tour guide robot (Burgard et al., 1998) is equipped with robust navigational skills, which are crucial to its role; however, humans still do not perceive such a robot as their partner but see it merely as a museum orientation tool. What we recognize as a partner is probably a robot that can develop various kinds of relationships with humans. This does not mean simply performing multiple tasks. Rather, we believe that it is important to establish interactive relationships first, and then the tasks and skills of partner robots will gradually emerge along with advancing technologies.

## 2. FIELD TRIAL

Field trials provide an important means of exploring the potential of partner robots. We need extended observations because social relationships develop over time. In our field trial, two humanoid robots that had various communicative behaviors interacted with children at an elementary school. The purpose of the trial was for the robots to play with the children and to communicate with them in English, thus improving the children's ability to speak English. We observed the children's reactions to the robots over the course of 2 weeks. To the best of our knowledge, ours was the first extended trial using interactive humanoid robots in an authentic social setting.

Our choice of a task for the robot was motivated by the generally poor English language ability of Japanese people. We believe a lack of motivation and opportunities to speak English is a major cause of this deficiency. According to Gardner and Lambert (1972), the two main reasons for learning a second language are instrumental motivation (e.g., to earn a degree or obtain a position) and integrative motivation (e.g., to understand a different culture or to befriend foreigners). Many children in elementary and junior high school lack motivation and do not recognize the importance and usefulness of English. In fact, children have no need to speak English in Japan. Although English teachers speak English during class, children speak Japanese outside of class. In their daily lives, they almost never encounter foreigners who do not speak Japanese. Therefore, many children are not motivated to study English.

## 3. METHOD

We performed the field trial at an elementary school affiliated with Wakayama University. Two identical humanoid robots were put in the open corridor near the first- and sixth-grade classrooms, and for 2 weeks the two robots interacted with first-grade students and sixth-grade students. The following subsections describe the method of the trial in more detail.

### 3.1. Setting and Participants

We carried out two sessions, one for first graders and the other for sixth graders. In general, there are six grades in a Japanese elementary school. This particular elementary school has three classes for each grade and about 40 students in each class. There were 119 first-grade students (6–7 years old; 59 boys and 60 girls) and 109 sixth-grade students (11–12 years old; 53 boys and 56 girls).

Figure 1 shows the three classrooms of the first grade. There are no walls between the classrooms and corridor, so that the corridor (called a workspace) is open to every first grader. The first graders' classrooms are located on the ground floor; the sixth graders' classrooms have the same layout as the first graders' and are located on the third floor.

### 3.2. The Robot and Its Behavior

The interactive humanoid robot we developed used a wireless ID tag system to identify different individuals. With visual, auditory, tactile, and wireless ID tag information, the robot took the initiative in interacting with children. For example, it called a child's name and initiated interaction after detecting the child from his or her ID tag. The robot could only recognize and speak English,

*Figure 1.* Elementary school where robots were installed for 2 weeks, showing the open space plan of the school and where the cameras were installed.

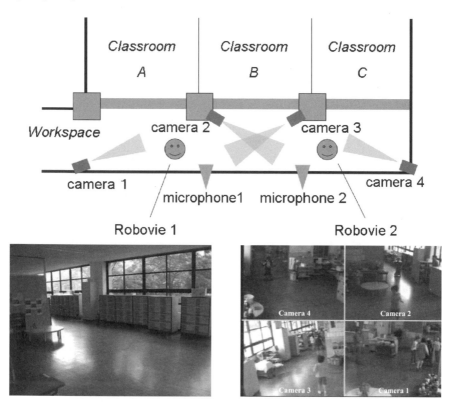

and its voice sounded somewhat like that of a child. The robot's utterances were based on recordings of a native English speaker (a professional narrator).

## Interactive Humanoid Robot "Robovie"

Figure 2 shows the humanoid robot Robovie (Ishiguro, Ono, Imai, & Maeda, 2001). The robot is capable of human-like expression and recognizing individuals by using various actuators and sensors. Its body possesses highly articulated arms, eyes, and a head, which were designed to produce sufficient gestures to communicate effectively with humans. The sensory equipment includes auditory, tactile, ultrasonic, and vision sensors, which allow the robot to behave autonomously and to interact with humans. All processing and control systems, such as the computer and motor control hardware, are located inside the robot's body.

*Figure 2.* Robovie and the wireless tag. Robovie (left) is an interactive humanoid robot that autonomously speaks, makes gestures, and moves around. With the antenna and tags, it is able to identify individuals.

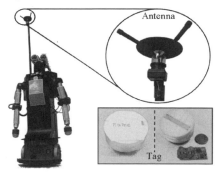

## Person Identification

To identify individuals, we developed a multiperson ID system for partner robots by using a wireless tag system. Recent radio frequency ID (RFID) technologies enabled us to use contactless ID cards in practical situations. In this study, children were given easy-to-wear nameplates (5 cm in diameter) in which a wireless tag was embedded. A tag (shown in Figure 2, lower right) periodically transmitted its ID to the reader, which was onboard the robot. In turn, the reader relayed received IDs to the robot's software system. It was possible to adjust the reception range of the receiver's tag in real time from software. The wireless tag system provided the robots with a robust means of identifying many children simultaneously. Consequently, the robots could show some human-like adaptation by recalling the history of interaction with a given person.

The robot could also distinguish between participants and listeners. In linguistic research, Clark (1996) classified people in the process of communicating into two categories: participants and listeners. Participants speak and listen, whereas listeners are an audience. Based on Clark's theory, we modeled daily communication among children and a robot as shown in Figure 3. This model does not include distant communication, such as a member of an audience questioning a presenter at a speech. The left side of the figure shows a situation in which a robot could not identify individuals. The right side of the figure shows the people around the robot classified into two categories: participants and observers. The participant category is similar to Clark's definition of participant, but the observer category does not include eavesdroppers who listen in without the speaker's knowledge, because we are only concerned with the people within the robot's sensor range. Furthermore, we assume that the distance between the robot and people is adequate for the robot to distinguish between the two categories.

*Figure 3.* The robot's communication model. The robot identifies multiple people simultaneously and classifies them into two categories for adapting its behaviors to them.

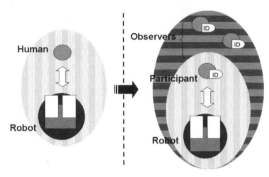

Hall (1966) discussed several zones of proximity between humans in a conversation. According to his theory, a conversational distance is within 1.2 m, and a common social distance for people who have just met is between 1.2 m and 3.5 m. In this study, we defined the participant as the person who was within 1.2 m and nearest to the robot. This definition is based on our assumption that the participant in the communication process would approach the robot as they interact. In addition, a previous study showed the average distance in human–robot interaction was about 50 cm (Kanda, Ishiguro, Ono, Imai, & Nakatsu, 2002), which also supports the contention that the participant will keep within 1.2 m. Meanwhile, other individuals who stayed within the detectable range of the robot were considered to be observers, because they did not communicate with the robot but were within its region of awareness. A detailed mechanism and performance of the person ID system is described in Kanda, Hirano, Eaton, and Ishiguro (2003).

## Interactive Behaviors

Robovie has a software mechanism for performing consistent interactive behaviors (Kanda, Ishiguro, Ono, Imai, & Mase, 2002). The intention behind the design of Robovie is that it should communicate at a young child's level. One hundred interactive behaviors have been developed. Seventy of them are interactive behaviors such as hugging (Figure 4), shaking hands, playing paper–scissors–rock, exercising, greeting, kissing, singing, briefly conversing, and pointing to an object in the surroundings. Twenty are idle behaviors such as scratching the head or folding the arms, and the remaining 10 are moving around behaviors. For the purpose of English education in this study, the situated module could only speak and recognize English. In total, the robot could utter more than 300 sentences and recognize about 50 words.

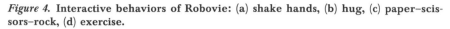
*Figure 4.* Interactive behaviors of Robovie: (a) shake hands, (b) hug, (c) paper–scissors–rock, (d) exercise.

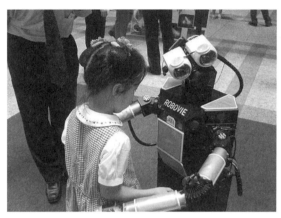

Several interactive behaviors depended on the person ID function. For example, there was an interactive behavior in which the robot called a child's name if that child was at a certain distance. This behavior was useful for encouraging the child to come and interact with the robot. Another interactive behavior was a body part game; the robot asked a child to touch a part of the body by saying the part's name.

These interactive behaviors appeared in the following manner based on simple rules. The robot sometimes triggered the interaction with a child by saying, "Let's play, touch me," and it exhibited idling or moving-around behaviors until the child responded; once the child reacted, it continued performing friendly behaviors as long as the child responded to it. When the child stopped reacting, the robot stopped the friendly behaviors, said, "good bye," and restarted its idling or moving-around behaviors.

## 3.3. Procedure

Both sessions (for first and sixth grade) were conducted for 2 weeks, which is equivalent to 9 days of school. We gave the children safety instructions before the trial. Pictures of the robot were accompanied by messages in Japanese such as, "Do not treat the robots roughly," and "Do not touch the joints because it is not safe." We did not give the children any further instructions.

The two robots were put in the corridor as shown in Figure 1 (indicated as Robovie 1 and Robovie 2). The children were allowed to interact freely with both robots during recess. Every child had a nameplate with an embedded wireless ID tag (Figure 2, right bottom) so that the robots could identify the child during interactions.

The teachers were not involved in the field trial. Two experimenters (university students) looked after the two robots. They did not help the children interact with the robots but simply ensured the safety of the children and robots. For example, when the children crowded closely around the robot, the experimenters would tell them to maintain a safe distance.

## 3.4. Data Collection

### Time Spent Interacting With the Robot

Each robot was equipped with a wireless ID tag reader that detected and identified ID tags embedded in the nameplates given to the children (described in Section 2). After identifying the children's IDs, the robot made a detection log of IDs for later analysis in addition to using it during interaction with the children. We prepared a simple program to calculate the interaction time per day for every child recorded in the detection log.

We also recorded scenes from the field trials with four cameras and two microphones. Figure 1 (upper and bottom right) describes the arrangement of the cameras and microphones and the obtained scenes of the trial. The video was used to verify the consistency of the wireless ID tag system. It was not analyzed otherwise.

### Tests of English Skills

The experimenters came to the first- and sixth-grade classes three times during the trial, and each time administered a brief English skills test: a pretest before the session, a test 1 week after the session began, and a posttest at the end of the 2-week session. Each test quizzed the students on the same six easy daily sentences used by the robots: "Hello," "Bye," "Shake hands please," "I love you," "Let's play together," and "This is the way I wash my face" (a phrase from a song), with the order of sentences changed in each test. We replayed the recorded voice of a native speaker for the test. On the answer sheets were four pictures for each phrase, and children had to choose the correct scene corresponding to the utterance (Figure 5). The score of the listening test for an individual was expressed as a percentage of the total number of correct answers, and thus the range of the listening test score always fell between 0 and 1.0.

### Social Interaction Around the Robot

We also administered a questionnaire that asked the children to write down the names of their friends. These names were compared with the log data from

*Figure 5.* Example of a question in the listening test: Which is "Bye"?

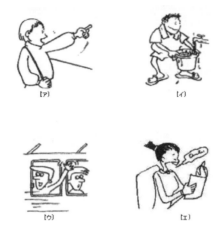

the wireless tags to calculate the time children spent with the robot and their friends together.

## 4. RESULTS

We analyzed the effect of the robots on social interaction over time and learning by conducting quantitative statistical tests on the tag data and the English test scores.

### 4.1. Preliminary Analyses

In Figure 6, we describe the main measurements used in this study (i.e., the number of minutes each child interacted with the robot in the 1st and 2nd weeks of the trial; their English scores on the pretest, 1st week test, posttest; and the amount and percentage of time they interacted with the robot in the presence of friends). The figure contains the correlations among the main variables. There was no overall improvement in English scores among the students (although, as noted later, we found improvement among those who spent more time with the robot in the 2nd week).

### 4.2. Grade Differences

In Figure 7, we show how the children in the first and sixth grades interacted with their robot over the 2-week period. The figure shows that first graders spent more time interacting with the robot than sixth graders did, and the

*Figure 6.* **Descriptive statistics and correlations among the measures.**

| Measures | M | SD | 1 | 2 | 3 | 4 | 5 | 6 |
|---|---|---|---|---|---|---|---|---|
| 1. Child's interaction time with robot in minutes, 1st week (1) | 12.5 | 14.0 | 1.00 | | | | | |
| 2. Child's interaction time with robot in minutes, 2nd week (2) | 2.7 | 5.4 | 0.27 | 1.00 | | | | |
| 3. Percentage interaction with friends (3) | 67% | — | −0.11 | −0.02 | 1.00 | | | |
| 4. Pretest English score | 0.69 | 0.16 | −0.02 | −0.11 | 0.08 | 1.00 | | |
| 5. English score after 1st week | 0.70 | 0.16 | −0.12 | −0.13 | 0.03 | 0.37 | 1.00 | |
| 6. English score after 2nd week | 0.69 | 0.16 | −0.04 | 0.10 | −0.05 | 0.35 | 0.40 | 1.00 |

*Note.* Correlations equal to or greater than ± .135 are significant at the .05 level or better.

*Figure 7.* **Interaction time with robots of first-grade students and sixth-grade students.**

| | 1st Week | | | | | 2nd Week | | | |
|---|---|---|---|---|---|---|---|---|---|
| Grade | 1 | 2 | 3 | 4 | 5 | 6 | 7 | 8 | 9 |
| 1st[a] | | | | | | | | | |
| M(min.) | 7.25 | 1.85 | 1.88 | 2.08 | 1.60 | 1.08 | 0.74 | 0.13 | 0.61 |
| SD | 7.36 | 3.57 | 3.14 | 4.90 | 3.77 | 3.00 | 2.43 | 0.51 | 2.35 |
| 6th[b] | | | | | | | | | |
| M(min.) | 3.33 | 3.09 | 0.59 | 1.15 | 1.30 | 1.31 | 0.79 | 0.20 | 0.77 |
| SD | 5.15 | 5.94 | 2.01 | 2.87 | 2.74 | 2.64 | 2.48 | 0.88 | 1.37 |

[a]$n = 119.$ [b]$n = 109.$

robot sustained their interest longer. It also indicates that the interaction between the children and the robots generally diminished in the 2nd week.

Nonetheless, a few children sustained a relationship with the robot. Child A said, "I feel pity for the robot because there are no other children playing with it," and Child B played with the robot for the same reason.

## 4.3. Social Interaction

We were surprised by the frequency with which children interacted with the robot in the company of other children (see Figures 8 and 9). Sixty-three percent of a first grader's interaction time with the robot was in the company of one or more friends. Seventy-two percent of a sixth grader's interaction time

*Figure 8.* Scenes of the interactions between Robovie and students. (a) First-grade students with the robot on Day 1. (b) First-grade students with the robot during the 2nd week. (c) Sixth-grade students on Day 1. (d) Sixth-grade students during the 2nd week.

(a)

(b)

(c)

(d)

*Figure 9.* **Transition in number of children playing with the robot. (a) Results for first-grade students. (b) Results for sixth-grade students. Number of interacting children represents the total number of the children identified by each robot's wireless system each day. Average of simultaneously interacting children represents the average number of children who simultaneously interacted with the robot. Rate of vacant time is the percentage of the time there was no child around the robot during each day.**

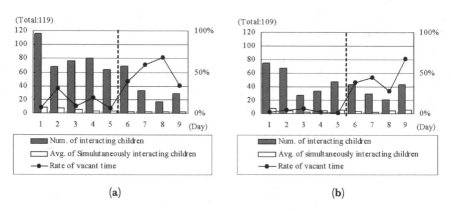

(a)                                    (b)

with the robot was in the company of one or more friends. Because the presence of friends could have affected each child's learning of English from the robot, we controlled for the presence of friends in the following analyses.

## 4.4. Learning English

The analyses we present are analyses of variance in which the dependent variable is the improvement in each child's English test score from the child's English pretest score. Although many children did not know English at the beginning of the trial, some knew a bit. If they knew any of the phrases on the English test (such as "bye") their improvement might have been small owing to a ceiling effect. Therefore, the appropriate analysis of the effects of the robot on learning is the change from the pretest to the posttest, controlling for the initial pretest score. The main analyses we ran were standard least squares analyses, described as follows:

Model (2nd week English score – pretest English score) = intercept + pretest English score + Week 1 interaction minutes with robot + Week 2 interaction minutes with robot + percentage of interaction time with friends.

The results of this analysis are shown in Figures 10 and 11. This analysis showed the expected significant ceiling effect of pretest English scores on the change in scores from pretest to posttest, $F(1, 198) = 86$, $p < .001$. That is, the more English the children already knew at the beginning of the trial, the less they learned from the robot. However, the amount of time they interacted

*Figure 10.* Analysis of variance results for effect of interaction time with the robot on improvement in English scores at the posttest (after 2 weeks).

| Source | df | F Ratio | p |
|---|---|---|---|
| Pretest English test score | 1 | 85.8 | < .0001 |
| Percentage of interaction time with friends | 1 | 1.5 | ns |
| Interaction time with robot, 1st week | 1 | 1.4 | ns |
| Interaction time with robot, 2nd week | 1 | 5.6 | .019 |
| Error | 198 | | |

*Figure 11.* Change in English score as a function of interaction time with the robot in the 2nd week, controlling for pretest English score and presence of friends.

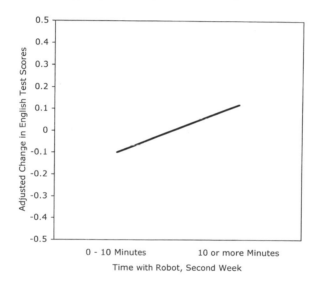

with friends and the robots together did not have an impact on the change in the English scores. The amount of time children spent with the robot during the 1st week also had no effect on their improvement in English by the 2nd week, but the amount of time that children interacted with the robots during the 2nd week did have a significant and positive impact on improvement in English in the 2nd week, $F(1, 198) = 5.6$, $p = .02$, $d = .33$.

Because we found significant improvement in English learning after 2 weeks, we examined whether there was any evidence of improvement after only 1 week with the robot. This analysis showed that time spent with the robot during the 1st week did not have a significant impact on the change in the English scores from the pretest to the 1st week's scores. Indeed, the trend was slightly negative $(p < .10)$. The absence of a 1st-week result suggests that learning depended on a sustained interest in the robot and maintaining a relationship with it. It was only those children who continued to interact with the robot through the 2nd week—those who formed a relationship with the robot—who learned from it.

We also investigated whether the grade of the children influenced the improvement in the English scores. To examine this we added the grade (first or sixth) to the previous equation, and included statistical interactions of grade with the presence of friends, time with the robot in the 1st week, and the presence of friends in the 2nd week. The results of this analysis did not change the overall positive effect of interaction time with the robot in the 2nd week (i.e., the relationship shown in Figure 10). However, this analysis did show that sixth graders learned more English than first graders ($p < .01$), and that first graders benefited slightly more from interaction with the robot in the 1st week ($p < .08$).

One alternative explanation to the improvement in English scores at the end of 2 weeks is that causality was reversed. That is, perhaps those children who were more interested in English and knew more English at the start of the trial were more interested in interacting with the robot. To investigate the possibility that knowledge of English caused the children to interact with the robot more, we ran a regression analysis examining the impact of pretest English scores on 1st and 2nd week of time spent with the robot, controlling for the presence of friends. Pretest English scores did not predict the 1st week of time with the robot, but there was a marginal positive effect of pretest scores on time with the robot in the 2nd week, $F(1, 208) = 2.5$, $p = .11$. This suggests that part of the reason for the results shown in Figures 10 and 11 might be the initial ability of some children to understand the robot's English and feel comfortable with it. They might have felt they had something in common with the robot (i.e., the English language).

## 5. DISCUSSION

We believe that this field trial provided us with many useful insights that we can apply to the development of future partner robots. The humanoid robots autonomously interacted with children by using their human-like bodies and various sensors such as visual, auditory, and tactile sensors. They also had a mechanism to identify individuals and to adapt their interactive behaviors to them.

The results suggest that the robot did encourage some children to improve their English, and that the robot was more successful in engaging children who already knew at least a little English. These findings support arguments based on previous literature in social psychology on similarity and common ground;

they suggest robots should be designed to have attributes and knowledge in common with their users.

## 5.1. Contributions to Human–Robot Interaction Methodology

Our results suggest that the impact of the robot did not show up until the 2nd week. This finding supports the argument that a robot's influence will depend on its ability to create a relationship with the user. It also suggests that a robot's effect on individuals changes over time. Therefore, we need to study long-term interactions to learn how to create effective partner robots.

## 5.2. Contributions to the Theory of Human–Robot Interaction

Our field trial highlighted the important unsolved aspects of human–robot interaction in an authentic social setting. The trial showed gradual loss of interest in interacting with the robot among most of the children. It was an important finding that the children interacted with the robot for the duration of 1 week; however, our robots failed to keep most of the children's interest after the 1st week. We believe that the robots' first impact created unreasonably high expectations in the children. The children mobbed the robot, overwhelming its ability to interact. In other words, the robot could not cope with the children's enthusiasm. Although partner robots are making news in Japan (such as Honda's and Sony's humanoid robots, and the big exhibition on partner robots named ROBODEX; ROBODEX Executive Committee, 2003), the robots' ability to be a partner to people is still lacking. Robots are very novel in general; therefore, their first impact can induce a greater desire for communication than their interactive ability can satisfy. In our trial, the children's interaction with the robots gradually decreased, especially during the 2nd week. Therefore, our trial showed us the limitation of the robots' ability to maintain long-term relationships and the disappointment that followed the robot's initial impact. However, we believe unreasonable expectations will diminish as partner robots become commonplace.

Regarding the body and appearance of the robot, our results seem to encourage the use of a humanoid robot. We believe that the body of a humanoid robot played a useful part in establishing common ground. That is, a robot that possesses a humanoid body will be more successful at sustaining interaction because people see it as similar to themselves and that it interacts as they do. Nonetheless, we need further research to establish a model of these kinds of social effects, such as common ground to see if they are more easily achieved with a humanoid robot by comparing humanoid and nonhumanoid robots.

It is also necessary to formalize a model of the relationships between humans and robots over time, and establish a method to promote lasting interactive relationships. Several pet robots have a special pseudolearning mechanism: Although they have many functions, they only show a few functions at first and then gradually reveal more according to their interactions. Furthermore, if robots really learn something about an individual to personalize the relationship, the robots will be able to build closer relationships with people. Therefore, identifying and defining the mechanism for sustaining long-term relationships is an important area of future research in human–robot interaction.

## 5.3. Contributions to the Design of Human–Robot Interaction

The trial showed that with respect to the interactive ability based on sensor data processing, real-world data are vastly different from that produced in a well-controlled laboratory. For example, many children ran around and spoke loudly to the robot; thus, its speech recognition was not effective in the classroom where the trial was carried out. To design robots that operate in real-world settings, we must consider how to make sensing more robust. Although many researchers and developers have been developing and improving sensing technologies, such as vision processing and speech recognition, robots still have weak ability compared to that of humans.

Fortunately, the wireless ID of persons worked well in our trial. We believe that one of the potentially promising approaches for acquiring interactive ability in the real world is to use environment-based sensors such as the wireless ID tags. In the trial, we observed several positive effects of the ID tags on the children's interaction with the robots:

- Child C did not seem to understand English at all. However, once she heard her name uttered by the robot, she became quite pleased and began interacting more frequently with the robot.
- Children D and E counted how many times the robot called their respective names. D's name was called more often, so D proudly told E that the robot preferred D.
- Child F passed by the robot. He did not intend to play with the robot, but because he saw Child G playing with the robot, he joined in.

These examples suggest that person ID was one of the triggers of the interaction and an essential behavior for continuous interaction.

Our robot currently recognizes only those who are around it. That is, even if the robot is faced with multiple parties, it does not distinguish the relationships among them. However, as the previous example indicates, relationships among people might affect the interaction. For example, a child may take a friend to the robot, or someone may take part in the interaction because a friend is playing with the robot. Therefore, we believe a partner robot should also recognize relationships between children (friendship, hostility, etc.).

## 5.4. Limitations

This study was a field trial rather than a true experiment with controls. For example, we did not compare the robot with an ordinary computer English teaching game. A detailed experiment might offer more precise and reliable results on the teaching of English to Japanese students. However, our main goal was not to teach English optimally but to learn how to create partnership in a robot. We believe that field trials in a frontier research area (e.g., partner robotics) are essential for developing the discipline. A field trial provides us with valuable information on the deficiencies in our approach, which is helpful to inspire future technological developments. We would be pleased if this work inspired rigorous research in the social aspects of human–robot interaction.

We did not associate videotaped interactions with tag data from each child. We believe this kind of fine-grained analysis would be particularly useful, for example, in checking the number of utterances of each child and in observing how each child initiated interaction with the robot. In future research, it would be very helpful to code the videotapes and thereby combine qualitative observations with tag data, which lack detailed information.

## 6. CONCLUSIONS

We performed a field trial for 2 weeks using interactive humanoid robots with first- and sixth-grade elementary school students. In the trial, the robots behaved as English peer tutors for Japanese students. The results suggest that the robot did encourage some children to improve their English. Our findings demonstrate the possibility of having interactive robots work in our daily life, although the benefits may be still too small to justify practical application. If the interactive robots were to acquire a more powerful ability to maintain relationships with humans, we would feel more confident in them serving various roles in our daily life in the immediate future. This result would encourage further robotics and human–computer interaction research related to sociality (e.g., theory of common ground), expression ability including humanoid control, sensory and recognition ability, and more metalevel communication mechanisms.

## NOTES

*Acknowledgments.* We thank the teachers and students at the elementary school for their eager participation and helpful suggestions. We also thank the reviewers for their advice and help on this article.

*Support.* This research was supported by TAO Japan.

*Authors' Present Addresses.* Takayuki Kanda, Intelligent Robotics and Communication Laboratories, ATR, 2-2-2 Hikaridai, Soraku-gun, Seika-cho, Kyoto, 619-0224, Japan. E-mail: kanda@atr.co.jp. Takayuki Hirano, Intelligent Robotics and Communication Laboratories, ATR, 2-2-2 Hikaridai, Soraku-gun, Seika-cho, Kyoto, 619-0224, Japan. E-mail: t-hirano@atr.co.jp. Daniel Eaton, Intelligent Robotics and Communication Laboratories, ATR, 2-2-2 Hikaridai, Soraku-gun, Seika-cho, Kyoto, 619-0224, Japan. E-mail: deaton@atr.co.jp. Hiroshi Ishiguro, Adaptive Machine Systems, Graduate school of Engineering, Osaka University, 2-1 Yamadagaoka, Suita, Osaka, Japan. E-mail: ishiguro@ams.eng.osaka-u.ac.jp.

*HCI Editorial Record.* First manuscript submitted to ACM *Transactions on Computer Human Interaction* on February 3, 2003. Transferred to *Human-Computer Interaction* on June 17, 2003. Revision received July 28, 2003. Accepted by Sara Kiesler and Pamela Hinds. Final manuscript received September 1, 2003. — *Editor*

## REFERENCES

Breazeal, C., & Scassellati, B. (1999). A context-dependent attention system for a social robot. *Proceedings of International Joint Conference on Artificial Intelligence.* San Francisco, CA: Kaufmann.

Burgard, W., Cremers, A. B., Fox, D., Hähnel, D., Lakemeyer, G., Schulz, D., et al. (1998). The interactive museum tour-guide robot. *Proceedings of 15th National Conference on Artificial Intelligence.* Menlo Park, CA, AAAI.

Cassell, J., Bickmore, T., Billinghurst, M., Campbell, L., Chang, K., Vilhjalmsson, H., et al. (1999). Embodiment in conversational interfaces: Rea. *Proceedings of the CHI 99 Conference on Human Factors in Computing Systems.* New York: ACM.

Clark, H. H. (1996). *Using language.* Cambridge, England: Cambridge University Press.

Cowley, S. J., & MacDorman, K. F. (1995). Simulating conversations: The communion game. *AI & Society, 9,* 116-137.

Dautenhahn, K., & Werry, I. (2002). A quantitative technique for analyzing robot-human interactions. *Proceedings of the IEEE/RSJ International Conference on Intelligent Robots and Systems.* New York: IEEE.

Fish, R. S., Kraut, R. E., Root, R. W., & Rice, R. E. (1992). Evaluating video as a technology for informal communication. *Proceedings of the CHI 92 Conference on Human Factors in Computing Systems.* New York: ACM.

Friedman, B., Kahn, P. H., Jr., & Hagman, J. (2003). Hardware companions?—What online AIBO discussion forums reveal about the human-robotic relationship.

*Proceedings of the CHI 2003 Conference on Human Factors in Computing Systems.* New York: ACM.

Fujita, M. (2001). AIBO: Towards the era of digital creatures. *International Journal of Robotics Research, 20,* 781–794.

Fussell, S., & Krauss, R. M. (1992). Coordination of knowledge in communication: Effects of speakers' assumptions about what others know. *Journal of Personality & Social Psychology, 62,* 378–391.

Gardner, R. C., & Lambert, W. E. (1972). *Attitude and motivation in second language learning.* Rowley, MA: Newbury House.

Hall, E. (1966). *The hidden dimension.* New York: Anchor.

Hinde, R. A. (1988). *Individuals, relationships and culture: Links between ethology and the social sciences.* Cambridge, England: Cambridge University Press.

Hirai, K., Hirose, M., Haikawa, Y., & Takenaka, T. (1998). The development of the Honda humanoid robot. *Proceedings of the IEEE International Conference on Robotics and Automation.* New York: IEEE.

Isaacs, E., & Clark, H. (1987). References in conversation between experts and novices. *Journal of Experimental Psychology: General, 116,* 26–37.

Isbister, K., Nakanishi, H., Ishida, T., & Nass, C. (2000). Helper agent: Designing an assistant for human–human interaction in a virtual meeting space. *Proceedings of CHI 2000 Conference on Human Factors in Computing Systems.* New York: ACM.

Ishiguro, H., Ono, T., Imai, M., Maeda, T., Nakatsu, R., & Kanda, T. (2001). Robovie: An interactive humanoid robot. *International Journal of Industrial Robots, 28,* 498–503.

Kanda, T., Hirano, T., Eaton, D., & Ishiguro, H. (2003). Person identification and interaction of social robots by using wireless tags. *Proceedings of the IEEE/RSJ International Conference on Intelligent Robots and Systems.* New York: IEEE.

Kanda, T., Ishiguro, H., & Ishida, T. (2001). Psychological analysis on human–robot interaction. *Proceedings of the IEEE International Conference on Robotics and Automation.* New York: IEEE.

Kanda, T., Ishiguro, H., Ono, T., Imai, M., & Mase, K. (2002). A constructive approach for developing interactive humanoid robots. *Proceedings of the IEEE/RSJ International Conference on Intelligent Robots and Systems.* New York: IEEE.

Kanda, T., Ishiguro, H., Ono, T., Imai, M., & Nakatsu, R. (2002). Development and evaluation of an interactive humanoid robot "Robovie." *Proceedings of the IEEE International Conference on Robotics and Automation.* New York: IEEE.

Murray, S. L., Holmes, J. G., Bellavia, G., Griffin, D. W., & Dolderman, D. (2002). Kindred spirits? The benefits of egocentrism in close relationships. *Journal of Personality & Social Psychology, 82,* 563–581.

Nakata, T., Sato T., & Mori, T. (1998). Expression of emotion and intention by robot body movement. *Proceedings of the IAS-5 International Conference on Intelligent Autonomous Systems.* Amsterdam, The Netherlands: IOS.

NEC Corporation. (2002). *Personal robot PaPeRo.* Retrieved September 1, 2003, from http://www.incx.nec.co.jp/robot/PaPeRo/english/p_index.html

Ono, T., & Imai, M. (2000). Reading a robot's mind: A model of utterance understanding based on the theory of mind mechanism. *Proceedings of 17th National Conference on Artificial Intelligence.* Menlo Park, CA: AAAI.

Ono, T., Ishiguro, H., & Imai, M. (2001). A model of embodied communications with gestures between humans and robots. *Proceedings of 23rd Annual Meeting of the Cognitive Science Society.* Mahwah, NJ: Lawrence Erlbaum Associates, Inc.

Petersen, M. G., Madsen, K. H., & Kjær, A. (2002). The usability of everyday technology: Emerging and fading opportunities. *ACM Transactions on Computer–Human Interaction, 9,* 74–105.

Quek, F., McNeill, D., Bryll, R., Duncan, S., Ma, X. F., Kirbas, C., et al. (2002). Multimodal human discourse: Gesture and speech. *ACM Transactions on Computer–Human Interaction, 9,* 171–193.

Reeves, B., & Nass, C. (1996). *The media equation.* Cambridge, England: Cambridge University Press.

ROBODEX Executive Committee. (2003). *ROBODEX2003 (robot dream exposition).* Retrieved September 1, 2003, from http://www.robodex.org/

Sakagami, Y., Watanabe, R., Aoyama, C., Matsunaga, S., Higaki, N., & Fujimura, K. (2002). The intelligent ASIMO: System overview and integration. *Proceedings of the IEEE/RSJ International Conference on Intelligent Robots and Systems.* New York: IEEE.

Thrun, S. (2004). Toward a framework for human–robot interaction. *Human–Computer Interaction, 19,* 9–24. [this special issue]

Wada, K., Shibata, T., Saito, T., & Tanie, K. (2002). Analysis of factors that bring mental effects to elderly people in robot assisted activity. *Proceedings of the IEEE/RSJ International Conference on Intelligent Robots and Systems.* New York: IEEE.

Weizenbaum, J. (1966). Eliza: A computer program for the study of natural language communication between man and machine. *Communications of the ACM, 9*(1), 36–45.

HUMAN-COMPUTER INTERACTION, 2004, Volume 19, pp. 85–116

# Moonlight in Miami: A Field Study of Human–Robot Interaction in the Context of an Urban Search and Rescue Disaster Response Training Exercise

Jennifer L. Burke, Robin R. Murphy,
Michael D. Coovert, and Dawn L. Riddle
*University of South Florida*

**Jenny Burke** is an industrial–organizational psychologist with an interest in human–robot interaction; she is a research associate in the Center for Robot-Assisted Search and Rescue, and in the Institute of Human Performance, Decision Making and Cybernetics, both located at the University of South Florida. **Robin Murphy** is active in the robotics research and applications communities; she is a professor in the Department of Computer Science & Engineering of the University of South Florida and Director of the Center for Robot-Assisted Search and Rescue. **Michael Coovert** is an industrial–organizational psychologist with an interest in technology's impact on individuals and organizations; he is a professor in the Department of Psychology of the University of South Florida and founder of the Institute of Human Performance, Decision Making & Cybernetics. **Dawn Riddle** is an industrial–organizational psychologist in the Department of Computer Science & Engineering of the University of South Florida.

## CONTENTS

## ABSTRACT

This article explores human–robot interaction during a 16-hr, high-fidelity urban search and rescue disaster response drill with teleoperated robots. This article examines operator situation awareness and technical search team interaction using communication analysis. It analyzes situation awareness, team communication, and the interaction of these constructs using a systematic coding scheme designed for this research. The findings indicate that operators spent significantly more time gathering information about the state of the robot and the state of the environment than they did navigating the robot. Operators had difficulty integrating the robot's view into their understanding of the search and rescue site. They compensated for this lack of situation awareness by communicating with team members at the site, attempting to gather information that would provide a more complete mental model of the site. They also worked with team members to develop search strategies. The article concludes with suggestions for design and future research.

# 1. INTRODUCTION

Urban search and rescue (USAR) involves the rescue of victims from the collapse of a man-made structure. The environment can be characterized as a pile of steel, concrete, dust, and debris. The areas are dark and perceptually disorienting; they no longer look like recognizable structures due to the collapse. Robot-assisted search and rescue in this field domain requires that physically situated robots operate under these unstructured, outdoor environmental conditions in areas that are either too narrow or unsafe for human or canine entry.

The relation between humans and robots in USAR is different than the relation in manufacturing, office, or even security applications of robots. Possibly, the most interesting human–robot interaction aspect of USAR is that robots, much like search dogs, must team with people to perform any activity. The teaming is physical because of the mobility challenges imposed by the USAR environment; robots must be carried in backpacks to the voids targeted to be searched and, once the robots are on the ground, they must be tethered. The teaming also is perceptual and cognitive because people must make decisions for the robots and interpret the video, audio, and thermal imaging data provided from the robots. The robots are short, providing a viewpoint from less than 1 ft off the ground. This viewpoint exacerbates a keyhole effect (i.e., the limited angular view associated with many remote vision platforms that gives remote observers a sense of trying to understand the environment through a peephole; Woods, Tittle, Feil, & Roesler, 2004). People must fuse these data with other data sources (e.g., building plans) and knowledge (e.g., time of day) to identify victims and structural anomalies as well as to conduct and coordinate the rescue efforts. Consequently, human–robot interaction in USAR requires distributed information transfer and cooperation. This task must be accomplished even while operators and decision makers (consumers of information provided by the robots) are under extreme cognitive and physical fatigue, introducing new issues not commonly seen in industrial settings. Furthermore, USAR is a domain where the robots perform tasks that cannot be accomplished by a living creature; thus, the operator has no higher metaphor or example of how to use the robot. The high degree of human involvement is not expected to change in the near future. The robots are not autonomously mobile for the demanding conditions of a rubble pile, and the most optimistic road map posits only navigational autonomy within 10 years (Murphy, 2002).

This article investigates human–robot interaction during robot-assisted search and rescue activities observed as part of a high-fidelity USAR field training drill in Miami, Florida, managed by Rescue Training Associates. It should be emphasized that data collection was opportunistic and observational: The drill was not structured for a formal human–robot interaction study, and we generated no hypotheses beforehand. Our observations suggested we carry out analysis using the perspectives of previous work in situation awareness and teamwork, which indicates need for effective task performance in complex, high-stress work domains similar to USAR (Prince & Salas, 2000; Sonnenwald & Pierce, 2000; Stout, Cannon-Bowers, Salas, & Milanovich, 1999), and human–robot interaction studies of USAR (Casper & Murphy, 2002, 2003) support the need for situation awareness. The analyses reported in this article focus on situation awareness and team process and communication in robot-assisted search and rescue.

## 2. OVERVIEW OF TECHNICAL SEARCH AND USAR

Technical search is one of many emergency response tasks. Its organizational structure poses interesting challenges for effective human–robot interaction. Here we summarize this structure (for more detail, see Casper, 2002).

In the United States, operations at a mass-casualty incident are divided into 12 emergency support functions, ranging from medical support to logistics. Each emergency support function is conducted by a specially trained task force and coordinated through an incident commander and the incident command staff. USAR is one function within the larger incident organization. Technical search, as seen in Figure 1, is one of the four USAR subspecialties within the task force: search, technical support, medical, and rescue or extrication. Although no two disasters are managed precisely the same way, USAR operations often begin with a manual reconnaissance of the area of damage, called the *hot zone*. Victims on the surface or easily removed from light rubble are extracted immediately as encountered. After reconnaissance, the command staff determines what the safest strategy is to effectively search the hot zone for survivors within the rubble. In areas that are deemed safe for people and dogs to investigate, canine teams may be sent forward.

USAR personnel require advanced training and equipment; this training is usually conducted by a designated federal or state task force. There are currently 28 federal task forces recognized by the Federal Emergency Management Agency and up to four times as many teams responsible for highly populated urban areas of states. Both federal and regional teams typically share the same organization, fielding a 56-person task force to sustain operations around the clock (in 12-hr shifts) for a maximum of 10 consecutive days. The teams are composed of firefighters, paramedics, and emergency medical technicians and civilians, most often in canine searches, structures, and hazardous materials tasks. USAR workers routinely log over 200 hr of USAR-specific training each year. Personnel who conduct technical searches are highly trained members of a cohesive team. They usually work in pairs for safety.

*Figure 1.* **Organizational structure of urban search and rescue task force (adapted from the United States Federal Emergency Management Agency, 1992).**

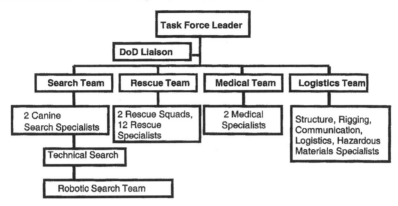

USAR operations are physically and cognitively fatiguing. Every member who works in the hot zone must be able to physically negotiate rubble piles, work in confined spaces, climb ladders and work at heights, and quickly exit void spaces to avoid secondary collapses. Task force members wear specialized safety equipment and are closely monitored for signs of physical exhaustion or stress (particularly critical incident stress syndrome). Although the teams work in 12-hr shifts, the reality of both shifts setting up operations and infrastructure and working in the field during the first 24 hr leads to sleep deprivation. It is conventional wisdom that a responder will get less than 3 hr of continuous sleep during the first 48 hr of an incident.

The command staff attempts to minimize the number of people in the hot zone, so technical search specialists wait at the forward station of the hot zone perimeter until called over the radio or assigned to an area to search. A technical search specialist may carry a fiber-optic boroscope, thermal imager, or a video camera mounted on a wand for a visual inspection of the rubble, depending on the verbal description of the void or the specific request of a particular device by the leader. If a survivor is found, the search team and command staff brings in the medical and rescue teams, who call on members of the technical support team as needed. Before leaving the void, the technical search specialists mark the exterior of the void with symbols indicating that it has been searched, the structural condition, and presence of survivors or remains.

## 3. RELATED WORK

Human–robot interaction is a comparatively new field (see Burke, Murphy, Rogers, Lumelsky, & Scholtz, 2004). Our study differs from existing research in human–robot interaction in three respects: goals, methodology, and focus.

Most human–robot interaction studies with physical robots have had the goal of understanding and improving social acceptance of robots or robot interface design (notably, Arkin, Fujita, Takagi, & Hasegawa, 2003; Breazeal, 2000; Draper, Pin, Rowe, & Jansen, 1999; Khatib, Yokoi, Brock, Chang, & Casal, 1999; Wilkes et al., 1999). By contrast, the goal of our study is to examine the interactions and relations between people and robots performing tasks in critical work contexts. Methods used by human–robot interaction researchers vary; experiments in laboratory or other controlled settings, simulations, and modeling are the most common methods for human–robot interaction studies. Comparatively few studies are conducted in actual task conditions where the main purpose is the work rather than data collection (e.g., Breazeal, 2003; Fong, Thorpe, & Baur, 2001; Kawamura, Nilas, Muguruma, Adams, & Zhou, 2003; Kiesler & Goetz, 2002; Langle & Worn, 2001; Severinson-Eklundh, Green, & Huttenrauch, 2003). By contrast, our method was to observe people deploying robots in collapsed buildings and rubble during a 16-hr training exercise. Finally, our work differs in focus from most previous human–robot interaction studies in that these studies have been robocentric (i.e., the robot's capabilities and experiences were the central point of interest). Our study focuses on the human side of the relation, exploring rescue workers' reactions and experiences as they worked with the robots.

There are three studies, however, that are relevant to this one: two directly in human–robot interaction for USAR (Casper & Murphy, 2002, 2003) and one in human–robot interaction for SWAT teams (H. Jones & Hinds, 2002). In each of these, situation awareness (Endsley, 1988) is a key construct for understanding (and improving) human–robot interaction. Just prior to the September 11, 2001 World Trade Center disaster, Casper and Murphy (2002) conducted an ethnographic study of Florida Task Force 3 members using robots to search for a victim (a fireman-down scenario in a partially collapsed building). The study showed that searching for victims and structural inspection, two of the activities described as part of the technical search task (see Section 2), are conducted simultaneously and suggested that two operators are needed to interpret multiple sensor data while navigating.

Casper and Murphy's (2003) analysis of video data collected during the World Trade Center disaster response found that a variety of human–robot interaction issues impacted performance of human–robot teams on the pile. The most relevant to this study were operators' lack of awareness regarding the state of the robot and the situatedness of the robot in the rubble. Operators also had difficulty linking current information obtained from the robot to existing knowledge or experience (see Casper, 2002). The Florida task force and World Trade Center human–robot interaction studies reveal difficulties in operator teleproprioception and telekinesthesis, consistent with the problems described in Sheridan (1992.)

In a domain very similar to USAR, H. Jones and Hinds (2002) observed police SWAT teams in training exercises and identified leader roles in establishing common ground and coordinating distributed team member actions as factors transferable to system design for coordinating distributed robots. Like search and rescue teams, SWAT teams operate in high-stress, time-critical work environments. Researchers found that leaders formed global mental models to build common ground (shared situation awareness) among distributed team members as they coordinated and directed distributed SWAT teams. Objects and spatial relations were used to coordinate actions with SWAT team members, and sharing common ground from the recipient's perspective increased situation awareness and team performance. The authors incorporated these findings into a system design using an object-centered electronic dialogue between an operator and multiple, distributed robots. They created a correspondence agent to assist the operator in building global situation awareness, and to send commands to distributed robots using their own frame of reference.

Our field study of team-based USAR operations differs from H. Jones and Hinds's (2002) work, which studied distributed SWAT teams (people) to model a team of distributed robots that could work together in a similar fashion (but not specifically with police SWAT teams). We observed real robot–user interaction as it occurred between team members and a single robot to inform the development of coordinated human–robot systems within the organizational structure of USAR. Their findings regarding the criticality of shared awareness in team-based, dynamic work domains, however, are applicable to our domain.

The findings from these studies all point to situation awareness, perception, and communication during tasks as critical aspects of human–robot interaction. Operators in field tests and at the World Trade Center did not know how to interpret what they saw through the robot's camera, partly because of fatigue and partly because of the lack of expected perceptual cues and the keyhole effect (Casper & Murphy, 2002, 2003.) Like the distributed team members in the SWAT team exercise, they needed to establish shared common ground in their own frame of reference (H. Jones & Hinds, 2002.)

The exploration of situation awareness in robot-assisted search operations in Section 5 is based on Endsley's (1988) three-level model, which defines situation awareness as "the perception of the elements in the environment within a volume of time and space, the comprehension of their meaning and the projection of their status in the near future" (p. 97). Perception (Level 1) is detection of sensory information: the perception of elements in the environment within a volume of time and space. Comprehension (Level 2) is divided into two subcategories: identification and interpretation. Identification is defined as comprehension of perceived cues in terms of subjective meaning (e.g., identifying objects, locations, and victims). Interpretation is defined as comprehension of perceived cues in terms of objective significance or importance to this situation. Projection (Level 3) is defined as the projection of future situation events and dynamics through projecting, generating, and activating solutions or plans.

Endsley's (1988) model is based on an information-processing theory (Wickens, 1992) in which situation awareness results from a process whose external source is primarily sensory information acquired through sight, sound, touch, taste, and smell. Perception and attention subprocesses take sensory data into working memory where they are coded and pattern matched with existing goals and mental models in long-term memory. D. Jones and Endsley (1996) noted that 76% of situation awareness errors in pilots were due to problems in perception (i.e., not noticing key information in the environment). This problem is of particular interest in our study, where the impact of perception on the control of robots is expected to be similarly problematic.

Situation awareness also requires other information and knowledge in addition to sensory input (e.g., system knowledge, prior knowledge, and the behavior of other people in the environment). Mental models play an important role in situation awareness. Operators develop internal representations of the technology they use and the environment in which they use it. These mental models help direct limited attention efficiently, integrate information, and provide a way of projecting future events or states. As Endsley (2000) stated, "the use of mental models in achieving situation awareness is considered to be dependent on the ability of the individual to pattern match between critical cues in the environment and elements in the mental model" (p. 16). Mental models support situation awareness; they can also hinder it if the mental models are inaccurate. In the D. Jones and Endsley (1996) study referenced earlier, 20% of situation awareness errors were associated with problems with mental models.

Research on teams and mental models has suggested that having a shared mental model of the problem space can increase situation awareness and team performance (H. Jones & Hinds, 2002; Sonnenwald & Pierce, 2000; Stout et al., 1999). Effective planning and communication strategies were found to increase team shared mental models, and correspondingly, team performance. In a study of military command and control exercises, Sonnenwald and Pierce found frequent communications between team members about the work context and situation; work process and domain-specific information were needed to maintain shared situation awareness in dynamic, constraint-bound contexts.

## 4. METHOD

This study was inductive and observational. We observed and videotaped five operators using one of three Inuktun robots in a 16-hr disaster response training drill. We later coded the operators' statements into categories generated by a content analysis of the operator statements.

## 4.1. Participants, Setting, and Procedure

### Setting

The setting for our study was a 3-day disaster response training exercise in Miami, Florida on November 28 through 30, 2001. The exercise consisted of 2 days of intensive hands-on training, which included collapse shoring, concrete breaching and breaking, heavy metal cutting and crane operations, technical search operations, and weapons of mass destruction–hazardous materials operations followed by a 16-hr deployment drill on an actual collapse site. As part of the Technical Search Operations module, which exposes course participants to the latest technical search innovations, all students received 20 min of awareness-level instruction in rescue robotics conducted by researchers from the Center for Robot-Assisted Search and Rescue. The awareness training course was designed to provide the students with a mental model of how the robot worked, and to provide an opportunity for hands-on experience teleoperating a robot (although time constraints precluded all students from having the chance to do so).

For the 16-hr high-fidelity response drill, a two-story warehouse in a light industrial park near the airport was partially collapsed, creating a large rubble pile. In addition to the collapsed building, two large rubble piles and an abandoned automobile that was set on fire were used for training operations. Figure 2 shows the layout of the collapse site and debris and rubble piles. The site was not simplified, and significant safety hazards were present. Large chunks of concrete walls, tangled rebar, and loose electrical wiring posed the main hazards to people on the piles. Weather and visibility conditions are not always conducive to rescue operations, but in this case, the night was clear (almost full moon) and the temperature normal for the area (70 °F).

In this kind of disaster environment, the visual technical search task consists of four activities in order of importance: search for signs of victims, report of findings to the team or task force leader, note any relevant structural information that might impact the further investigation of the void, and estimate the volume that has been searched and map it relative to the rubble pile. The technical search task is highly focused and generally limited to a short period of time where the searcher is called onto the pile, carries the technical equipment to the site, sets it up, gets results, and then returns to the forward station. In the Miami drill, robots collected this visual data. In our study, we attempted to capture how the operator used the robot to search for signs of survivors and

94

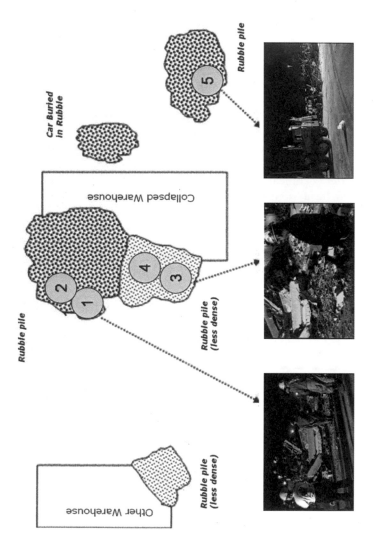

*Figure 2.* Map of disaster response training site and robot run locations.

noted structural information, because these were the activities with direct human–robot interaction.

## Participants

The 5 participants in the study were 3 student participants of the disaster response training exercise and 2 instructors. These participants were a subset of the approximately 75 students and approximately 15 instructors involved in the drill, who can be characterized as current USAR task force members serving as instructors or completing required recertification training hours, or first responders (firefighters and emergency medical technicians) seeking USAR certification to be eligible to serve on a regional task force team. The majority of students had no previous USAR experience in a large-scale disaster, such as the collapse of a large building due to an explosion or earthquake.

## Apparatus

The apparatus used in the study consisted of three robotic systems: two Inuktun Microtracs System robots (see Figure 3a) and an Inuktun Micro Variable Geometry Tracked Vehicle robot (see Figure 3b). The user interface offers little information beyond a visual view of the environment from the robot's camera. Scale, dimensionality, and color resolution are known constraints. The three robots are small, tracked platforms equipped with a color CCD camera on a tilt unit and two-way audio through a set of microphones and speakers on the robot and operator control unit. The Inuktun Micro Variable Geometry Tracked Vehicle is a polymorphic robot that can change from a flat position to a raised, triangular position. Its design allows the vehicle to change shape while moving to meet terrain and obstacle challenges, and it is capable of lifting the camera up to a higher vantage point (about 10.5-in. high when raised to maximum height). All three robots are powered and controlled through a 100-ft tether cord that connects the operator control unit and the robot. The Inuktun robots have limited communication capability. The operator is given basic control capability: traversal, power, camera tilt, focus, illumination, and height change for the polymorphic robot.

## Procedure

At the start of the drill, participants were checked in, divided into three teams, assigned roles, and transported to the site. Once at the site, they established scene security, set up the base of operations, and conducted site safety and operational surveys. Field operations commenced at 10:30 p.m., approxi-

*Figure 3.* (a) **Inuktun Microtracs, (b) Variable Geometry Tracked Vehicle robots.**

(a)

(b)

mately 4 hr after the drill began. During field operations, the robot cache was available for deployment on call. Robots were deployed in three areas of the hot zone, as shown in Figure 2. When a team requested a robot via radio, two or three researchers would move to the requested location and set up the robot for use, explaining the controls to the operator as needed. A student or researcher was designated as tether manager for the operator (i.e., uncoiled and recoiled the tether cord, and sometimes shook or popped the cord to free it from debris).

Our data collection was a modified version of the procedure used by Casper (2002). Two cameras simultaneously recorded the view through the robot's camera (what it sees) and a view of the operator and the operator control unit (what the operator is seeing and doing.) When the robot was visible, a third video unit recorded an external view of the robot in use.

The robot was deployed five times (see Figures 2 and 4). Three of the five runs (Runs 1, 2, and 5) were initiated on request by the teams; the other two runs

*Figure 4.* Operator run times, robots used, statement frequencies, and ratios

| Operator No. | Start Time (Approximate) | Duration (Min:Sec) | Robot Used | Human–Robot Ratio | Total No. Operator Statements | Statements: Minute Ratio |
|---|---|---|---|---|---|---|
| 1(S) | 10:45 p.m. | 14:20 | VGTV | 3:1 | 82 | 5.73:1 |
| 2(S) | 11:25 p.m. | 13:48 | VGTV | 2:1 | 66 | 4.78:1 |
| 3(I) | 12:45 p.m. | 14:39 | VGTV | 3:1 | 54 | 3.68:1 |
| 4(I) | 1:05 a.m. | 14:52 | VGTV | 2:1 | 60 | 4.03:1 |
| 5(S) | 3:15 a.m. | 3:42 | Micro-Tracs | 2:1 | 10 | 2.70:1 |
| M |  | 12:16 |  |  | 54 | 4.4:1 |
| SD |  | 4:48 |  |  | 24 | 1.17:1 |

*Note.* S = student; I = instructor; VGTV = Variable Geometry Tracked Vehicle.

(Runs 3 and 4) were initiated by instructors to gain hands-on experience with the robots. The first two runs searched the main rubble pile located next to the collapsed building. In the next two runs, areas that had already been searched by the teams were explored. The fifth run used the robot during victim recovery operations on the smaller rubble pile in an attempt to get a visual of or pathway to the victim. In each run, members of the team self-organized to use the robot. In Runs 2, 4, and 5, two members of the team were involved in robot operation. In Runs 1 and 3, a third participant got involved spontaneously by looking over the shoulder of the operator and interacting with the operator. The remainder of the team was either occupied with other tasks or passively observed. The five runs yielded 66 min, 16 sec of videotape for analysis.

## 4.2. Measures

As we observed the operators throughout the course of the drill, we realized that they did not work alone, and that the communications between team members played an important role in operator behaviors and performance. Therefore, we sought to measure communication between the operator and other team members as well as situation awareness. Because robot-assisted search and rescue is a relatively new field, there are no existing domain-relevant methods of analysis (e.g., communication coding schemes). The Federal Aviation Administration's Controller-to-Controller Communication and Coordination Taxonomy (C⁴T; Peterson, Bailey, & Willems, 2001) uses verbal information to assess team member interaction from communication exchanges in an air traffic control environment. The C⁴T is applicable to our work in that it captures the "how" and "what" of team communication by coding form, content, and mode of communication. Our goal, however, was twofold, not only to capture the how and what of USAR robot operator teams, but also the "who," and to capture observable indicators of robot operator situational awareness. Therefore, we developed a new coding scheme, the Robot-Assisted Search and Rescue Communication Coding Scheme (RASAR CCS). Although the development of the coding scheme is guided by the structure of the C⁴T, and incorporates relevant portions of the C⁴T, the RASAR CCS is domain specific. It was developed to examine USAR robot operator interactions with team members and to capture observable indicators of robot operator situation awareness.

The RASAR CCS addresses our goals of capturing team process and situation awareness by coding each statement on four categories: (a) speaker–recipient dyad, (b) form or grammatical structure of the communication, (c) function or intent of the communication, and (d) content or topic of the communication. By examining dyad, form, and content, we can determine which team members are interacting and what they are communicating about. Similarly, exploring elements of content and function allows us to examine indicators of operator situation awareness. The development of the RASAR CCS is described later, and the complete coding scheme is provided in Figure 5.

**Dyads**

Speaker–recipient dyad codes were developed as a function of speaker–recipient pairs of individuals anticipated in a USAR environment. Nine dyads were constructed to describe conversations between individuals. Five dyad codes classify statements made by the operator to another person (or persons): the tether manager, another team member, the researcher or robot technician, group, or other. The remaining four classify statements received by the operator from another person: the tether manager, another team member, the researcher or robot technician, or other.

The primary dyads involve the operator and tether manager (the person manipulating the robot's tether during teleoperation), operator and researcher, or operator and another team member. The element *operator–other* is used when the operator addresses a specific person who does not match one of those roles. The *operator–group* dyad is used when the operator is addressing those present as a group, or when the operator's statements are not clearly addressed to a specific individual. Verbalizations between individuals that did not include the operator were not coded.

**Form**

The form category contains the following elements: question, instruction, comment, or answer. (A statement can be a whole sentence or a meaningful

*Figure 5.* **Robot-Assisted Search and Rescue Communication Coding Scheme.**

| Category | Subcategories | Definitions |
|---|---|---|
| Sender–recipient dyad | Operator–tether manager | Operator: Individual teleoperating the robot |
| | Tether manager–operator | Tether manager: Individual manipulating the tether and assisting operator with robot |
| | Team member–operator | Team member: One other than the tether manager who is assisting the operator (usually by interpreting) |
| | Operator–team member | |
| | Researcher–operator | Researcher: Individual acting as scientist or robot specialist |
| | Operator–researcher | |
| | Other–operator | Other–individual interacting with the operator who is not a tether manager, team member, or researcher |
| | Operator–other | |
| | Operator–group | Group–set of individuals interacting with the operator |
| Statement form | Question | Request for information |
| | Instruction | Direction for task performance |
| | Comment | General statement, initiated or responsive, that is not a question, instruction, or answer |
| | Answer | Response to a question or an instruction |
| Content | State of the robot | Robot functions, parts, errors, capabilities, and so forth |
| | State of the environment | Characteristics, conditions, or events in the search environment |
| | State of information gathered | Connections between current observation and prior observations or knowledge |
| | Robot situatedness | Robot's location and spatial orientation in the environment; position |
| | Victim | Pertaining to a victim or possible victim |
| | Navigation | Direction of movement or route |
| | Search strategy | Search task plans, procedures, or decisions |
| | Off task | Unrelated or extraneous subject |
| Function | Nonoperator | Default for statements made by individuals other than the operator |
| | Seek information | Asking for information from someone |
| | Report | Sharing observations about the robot, environment, or victim |
| | Clarify | Making a previous statement or observation more precise |
| | Confirm | Affirming a previous statement or observation |
| | Convey uncertainty | Expressing doubt, disorientation, or loss of confidence in a state or observation |
| | Plan | Projecting future goals or steps to goals |
| | Provide information | Sharing information other than that described in report, either in response to a question or offering unsolicited information |

99

phrase or sentence fragment.) Statements not matching these categories are classified as *undetermined*.

To establish content and function codes, a subset of operator statements (177 of the 272 total statements) were subjected to a Q-sort content analysis (Sachs, 2000). Two subject matter experts not involved in the study sorted operator statement on content (according to the topic being discussed) and on function (according to the high-level purpose of the statement). Q-sort categories were reviewed and refined by two additional subject matter experts to ensure the elements reflected the domain of content and function.

## Content

The Q-sort analysis based on content yielded seven elements representing the content category:

1. Statements related to robot functions, parts, errors, or capabilities (state of the robot).
2. Statements describing characteristics, conditions, or events in the search environment (state of the environment).
3. Statements reflecting associations between current observations and prior observations or knowledge (state of information gathered).
4. Statements surrounding the robot's location, spatial orientation in the environment, or position (robot situatedness).
5. Indicators of direction of movement or route (navigation).
6. Statements reflecting search task plans, procedures, or decisions (search strategy).
7. Statements unrelated to the task (off task).

The first four content elements are necessary for building and maintaining situation awareness in search operations, whereas the elements of navigation and search strategy require situation awareness. Situation awareness is generated through information perceived (Level 1) and comprehended (Level 2) about the robot and environment. Because navigation and search strategy are elements that cannot be executed efficiently without situation awareness, statements reflecting these are indicators of operator situation awareness (Level 3).

## Function

Eight elements were identified from the Q-sort to represent the function category:

1. Asking for information from someone (seek information).
2. Sharing observations about the robot or environment (report).
3. Making a previous statement or observation more precise (clarify).
4. Affirming a previous statement or observation (confirm).
5. Expressing doubt, disorientation, or loss of confidence in a state or observation (convey uncertainty).
6. Projecting future goals or steps to goals (plan).
7. Sharing information other than that described in the report, either in response to a question or offering unsolicited information (provide information).
8. For this study, the focus is on operator situation awareness; hence, an eighth element was included as a default for statements made by individuals other than the operator (nonoperator).

The function elements of reporting and providing information merit explanation, as they appear very similar. Reporting involves perception and comprehension of the state of the robot, robot situatedness, the environment, or the state of information gathered. Any other information shared by an operator, in answer to a question or on his or her own, is classified as providing information (e.g., search strategy or navigation). Indicators of situation awareness are captured in the function category primarily through the elements of reporting and planning. When an operator shares information (reports) based on the robot's eye view, we can infer the first two levels of situation awareness, perception and comprehension, have taken place. The third situation awareness level, planning and projection, is captured in the function category as the element plan.

Raters also provided an overall assessment of each robot operator's situation awareness during the run, rated on a 5-point Likert scale ranging from 1 (*low*) to 5 (*high*). This observer rating is a subjective measure reached by consensus between the two raters, based on the following questions: How well did the operator understand what he or she was seeing through the robot's eye view, what the robot's state was at any given moment, and how the robot-supplied information related to other operator knowledge concerning the technical search operation? Video recordings of the operators manipulating the robot were used to code statements made by both the operators and surrounding personnel.

Two raters were trained to code the videotapes. One rater was Jennifer L. Burke, who was involved in data collection. The second rater, although not naive, was not on site during data collection. Raters reviewed descriptions of the disaster drill and data collection procedures, and then reviewed definitions for all the codes. Coding guidelines were developed to reduce ambiguity and to enhance reliability. Behavioral examples selected from the videotapes were also reviewed. The majority of the training centered on coding statements together and reaching consensus. Training continued until both raters felt comfortable rating independently (approximately 8 hr).

A written transcript of each videotape was produced, yielding a fixed number of statements to be coded (502 statements across the five operators). Using the Noldus Observer Video-Pro (Noldus, Trienes, Hendriksen, Jansen, & Jansen, 2000) observational coding software, raters coded 181 statements (36%) in the transcripts along four dimensions: dyad (speaker–recipient pair), form (grammatical structure of the communication), function (intent of the communication), and content (topic). Cohen's kappa (1960) was computed to measure interrater agreement for each of the four coding dimensions: dyad, form, function, and content. Reliability analyses verified modest reliability at better-than-chance levels, with Cohen's kappas of .72 for dyad, .78 for statement type, .64 for statement content, and .72 for statement function. The remaining 64% of statements were coded by a single rater.

Codes for each of the 502 statements were used in data analyses. Frequencies, percentages, and correlations of the coding categories and elements are generated to explore team process and communication: who talked with whom (dyad), how (form), about what (content), and for what purpose (function). This is an exploratory study in that we are looking for relations that may have some bearing on effective human–robot interaction in the USAR domain. Therefore, all operator statement categories were included in analyses. Significant relations emerged and are presented in each of the four categories. All correlations reported are significant at $p < .05$ unless otherwise noted.

## 5. RESULTS

### 5.1. Situation Awareness

Operators had difficulty building or maintaining situation awareness, and spent over one half of their time trying to do so. Figure 6 summarizes the percentage of statements the operator made in each category—sender–recipient dyad, function, form, and content. As shown in Figure 6d, 54% of operator statements were related to gaining situation awareness at Levels 1 and 2 (state of robot and robot situatedness, 38%; state of environment, 13%; and information gathered, 3%); considerably less time was spent talking about factors requiring situation awareness (Level 3) to perform (navigation, 21%; search strategy, 16%).

Relations between elements in the dimensions of content and function captured indicators of operator situation awareness. The correlation matrix of operator statement categories (Figure 7) revealed operator statements related to search strategy were strongly correlated with statements related to the state of

*Figure 6.* Coding category element frequencies and percentages for operator statements.

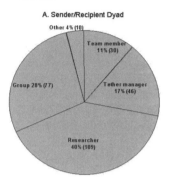

A. Sender/Recipient Dyad

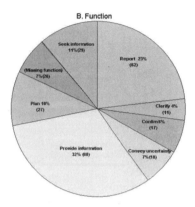

B. Function

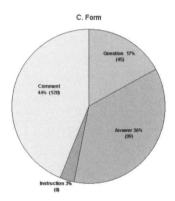

C. Form

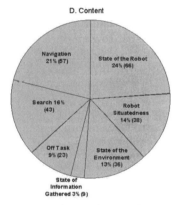

D. Content

the environment ($r = .94$) and state of information gathered ($r = .89$). These two situation awareness-related content areas were closely tied to each other ($r = .91$) as well, indicating the importance of linking what is being observed in the environment with what the operator already knows about the environment.

Search strategy and planning are an intuitive fit because of the need to plan search activities, and indeed, search strategy statements correlated with statements coded as planning ($r = .95$) in the function category. However, the significant correlation of planning (a situation awareness Level-3 indicator) with the state of the environment ($r = .98$, $p < .001$) emphasizes the necessity of perception and comprehension in performing search operations. This requirement is confirmed by another important relation in this category between the two functions of plan and report ($r = .93$.) The report element is used strictly when the operator is reporting on the state of the robot (including

*Figure 7.* Intercorrelations of operators' statements by coding category elements.

| Operator Statement Categories | 1 | 2 | 3 | 4 | 5 | 6 | 7 | 8 | 9 | 10 | 11 | 12 | 13 | 14 | 15 | 16 | 17 | 18 | 19 | 20 | 21 | 22 | 23 |
|---|---|---|---|---|---|---|---|---|---|---|---|---|---|---|---|---|---|---|---|---|---|---|---|
| Sender-recipient dyad | | | | | | | | | | | | | | | | | | | | | | | |
| 1. Operator-tether | 1.00 | | | | | | | | | | | | | | | | | | | | | | |
| 2. Operator-researcher | .03 | 1.00 | | | | | | | | | | | | | | | | | | | | | |
| 3. Operator-team member | −.26 | .35 | 1.00 | | | | | | | | | | | | | | | | | | | | |
| 4. Operator-other | .02 | .04 | .20 | 1.00 | | | | | | | | | | | | | | | | | | | |
| 5. Operator-group | .19 | .68 | −.03 | .63 | 1.00 | | | | | | | | | | | | | | | | | | |
| 6. Question | .14 | .82 | .59 | −.28 | .23 | 1.00 | | | | | | | | | | | | | | | | | |
| Statement form | | | | | | | | | | | | | | | | | | | | | | | |
| 7. Instruction | −.15 | .43 | .97** | .41 | .22 | .57 | 1.00 | | | | | | | | | | | | | | | | |
| 8. Answer | .31 | .89* | .08 | −.23 | .55 | .83 | .15 | 1.00 | | | | | | | | | | | | | | | |
| 9. Comment | .45 | .41 | .28 | .80 | .80 | .22 | .52 | .30 | 1.00 | | | | | | | | | | | | | | |

| | 1 | 2 | 3 | 4 | 5 | 6 | 7 | 8 | 9 | 10 | 11 | 12 | 13 | 14 | 15 | 16 | 17 | 18 | 19 | 20 | 21 | 22 | 23 |
|---|---|---|---|---|---|---|---|---|---|---|---|---|---|---|---|---|---|---|---|---|---|---|---|
| Content | | | | | | | | | | | | | | | | | | | | | | | |
| 10. State of the robot | .65 | .58 | −.38 | −.15 | .57 | .45 | −.26 | .84 | .35 | 1.00 | | | | | | | | | | | | | |
| 11. State of environment | −.22 | .62 | .87 | .50 | .46 | .59 | .95* | .30 | .61 | −.12 | 1.00 | | | | | | | | | | | | |
| 12. State of information | −.30 | .63 | .94* | .10 | .16 | .77 | .92* | .37 | .28 | −.15 | .91* | 1.00 | | | | | | | | | | | |
| 13. Robot situatedness | .69 | .53 | .42 | .00 | .31 | .77 | .49 | .67 | .57 | .58 | .44 | .47 | 1.00 | | | | | | | | | | |
| 14. Search | −.04 | .44 | .94* | .44 | .26 | .58 | .99** | .19 | .60 | −.18 | .94* | .89* | .57 | 1.00 | | | | | | | | | |
| 15. Navigation | .56 | .30 | −.45 | .53 | .84 | −.08 | −.20 | .39 | .71 | .71 | −.01 | −.34 | .28 | −.12 | 1.00 | | | | | | | | |
| 16. Off task | −.04 | .76 | −.27 | −.37 | .46 | .51 | −.24 | .84 | .72 | −.07 | −.02 | .08 | .16 | −.25 | .31 | 1.00 | | | | | | | |
| Function | | | | | | | | | | | | | | | | | | | | | | | |
| 17. Seek information | .28 | .75 | .46 | −.39 | .17 | .98** | .44 | .85 | .16 | .54 | .43 | .65 | .81 | .46 | −.05 | .53 | 1.00 | | | | | | |
| 18. Report | .35 | .68 | .66 | .43 | .60 | .71 | .80 | .57 | .82 | .35 | .83 | .72 | .84 | .85 | .34 | .09 | .65 | 1.00 | | | | | |
| 19. Clarify | .28 | .65 | −.09 | −.66 | .13 | .74 | −.13 | .88* | −.15 | .76 | −.06 | .19 | .52 | −.11 | .09 | .83 | .82 | .22 | 1.00 | | | | |
| 20. Confirm | .72 | .44 | .06 | −.45 | .09 | .72 | .06 | .76 | .17 | .75 | .17 | .18 | .86 | .13 | .20 | .41 | .84 | .48 | .81 | 1.00 | | | |
| 21. Convey uncertainty | .26 | .17 | −.49 | .67 | .81 | −.34 | −.25 | .13 | .62 | .44 | −.03 | −.41 | −.07 | −.19 | .93* | .21 | .14 | −.36 | −.19 | −.18 | 1.00 | | |
| 22. Provide information | .52 | .49 | .31 | .70 | .79 | .35 | .53 | .42 | .99* | .44 | .61 | .33 | .69 | .62 | .70 | .01 | .30 | .87 | −.01 | .32 | .56 | 1.00 | |
| 23. Plan | −.01 | .65 | .83 | .52 | .53 | .64 | .94* | .39 | .73 | .04 | .98** | .86 | .60 | .95* | .13 | .00 | .51 | .93* | .02 | .18 | .04 | .75 | 1.00 |

*$p < .05$. **$p < .001$.

105

situatedness), environment or information gathered, all indicators of perception, and comprehension (Levels 1 & 2 situation awareness.) The statistical association clearly ties situation awareness to operator planning (Level 3) in human–robot interaction.

## 5.2. Team Process and Communication

Operators demonstrated team-based processes and communication techniques while using the robot in search operations, a finding supported by statement frequencies, percentages, and correlations between statement categories. Results are first presented for the 272 statements made by the operators to team members, because the study's focus is on the operator's mental model and situation awareness. Additional results examining operator and team member statements are then presented. Figure 6a provides frequency and percentage of occurrence of each descriptor by coding category. As shown in Figure 4, operators spoke to other participants approximately four times per min while teleoperating the robot ($M = 4.4$, $SD = 1.17$ statements per minute).

Correlations of operator statement form with content (Figure 7) suggest operators' instructions were related to search strategy, the state of the environment, and state of information gathered ($rs = .99, .95, .92$, respectively.) In addition, instruction statements made by operators correlated significantly with statements coded as having a planning function ($r = .94$.) This result suggests operators were attempting to develop a shared mental model with teammates to increase situation awareness. They also used this information to plan and devise search strategies. The report function used in the coding scheme was defined as "reporting about the state of the robot, environment, or information gathered"—all situation awareness-related topics.

Our results show that reporting and planning were closely related (i.e., operators were using what they were seeing through the robot's eye to form a mental model of the search space, and the robot's position in that space, to devise search strategies). Planning not only facilitates the building of shared mental models with teammates, it also can result in improved team performance (Stout et al., 1999). It is surprising that navigation statements correlated only with statements function coded as conveying uncertainty ($r = .93$). This finding may reflect the lack of situation awareness in two of the operators.

Although the primary focus of this article is on operator situation awareness and how operators talk to team members to facilitate situation awareness, we did further analyses to explore information exchange between dyads on the team. That is, we examined operator statements both to and from primary rescue team members (operator, tether manager, team member, and researcher—robot specialist). In this analysis, we examine (by dyad) the frequency of statements based on form, content, and function combined to give an integrated picture of information exchange between rescue team members (e.g., the operator asked the tether manager a question seeking information about the state of the robot).

Naturally, at this level of detail, the number of possible combinations (four forms, seven topics, and seven functions) is formidable. Therefore, Figure 8 presents only the three highest frequency statement types (including ties) for each dyad; the percentages shown do not sum to 100%. Operator statements reflect specific expectations regarding the nature of each team member's roles (see Figure 9). The data suggest team members did not share the operator's role expectations. For example, although team members provided information on the robot and the environment, and provided instructions for navigation, they paid little attention to search strategy. In addition, tether managers provided information on the robot and its situatedness; however, they mainly provided instructions regarding navigation. This suggests operators saw tether managers as a resource for obtaining information, whereas tether managers saw their role as providing assistance with navigation. Although the operator saw team members as problem holders, sharing pertinent information about the state of the robot and the environment, and collaborating on search strategy, team members did not address operator needs regarding search strategy.

## 5.3. Interaction of Situation Awareness and Team Communication

As mentioned previously, the RASAR CCS obtains global assessments of situation awareness for each operator on a 5-point scale ranging from 1 (*low*) to 5 (*high*). These ratings were used to identify operators with high versus low situation awareness. Data from two operators receiving a rating of 1 were combined to form a low situation awareness group, and similarly, data from operators receiving a 4 or 5 were combined to form a high situation awareness group (data from one operator receiving a 3 were not used in this analysis). Chi-square analyses are computed to determine differences in high and low situation awareness operator statements relative to who the operator was communicating with (dyad), the statement form, content, and function.

Comparisons between operators rated as having high versus low situation awareness on a global rating scale offer support for the influence of team behaviors on situation awareness. Chi-square results (Figure 10) suggest operator communication with the tether manager, $\chi^2(1, N = 2) = 16.2$, $p < .001$; and with other team members, $\chi^2(1, N = 2) = 18.6$, $p < .001$; was related to high situation awareness. High situation awareness operators also provided instructions more frequently than their low situation awareness counterparts, $\chi^2(1, N = 2) = 4.5$, $p < .05$.

*Figure 8.* **Dyad frequencies and percentages for tether managers and team members.**

| Statement Type | Percentage of Speakers' Statements | Statement Type | Percentage of Speakers' Statements |
|---|---|---|---|
| Operator–tether manager exchanges $(n = 83)$ | | | |
| Operator to tether manager $(n = 47)$ | | Tether manager to operator $(n = 36)$ | |
| Question seeking information about the state of the robot | 9 | Instructions regarding navigation | 22 |
| Question seeking information about robot situatedness | 6 | Comment on robot situatedness | 11 |
| Question seeking information about navigation | 6 | Comment on navigation | 11 |
| Instruction planning the state of the robot | 6 | Answer about the state of the robot | 11 |
| Comment providing information on the state of the robot | 6 | | |
| Answer reporting navigation | 6 | | |
| Answer confirming navigation | 6 | | |
| Operator–team member exchanges $(n = 76)$ | | | |
| Operator to team member $(n = 27)$ | | Team member to operator $(n = 49)$ | |
| Question seeking information about the state of the robot | 7 | Comment on the state of the robot | 14 |
| Question seeking information about the state of the environment | 7 | Instruction on navigation | 12 |
| Question seeking information about search strategy | 7 | Comment on the state of the environment | 10 |
| Comment reporting the state of the environment | 7 | | |
| Comment reporting robot situatedness | 7 | | |
| Comment planning search strategy | 7 | | |
| Answer providing information about search strategy | 7 | | |

Furthermore, chi-square reveals that regardless of who they were speaking to, high situation awareness operators made more statements than low situation awareness operators about robot situatedness, $\chi^2(1, N = 2) = 5.4$, $p < .05$, and about search strategy, $\chi^2(1, N = 2) = 12.9$, $p < .001$. This finding suggests high situation awareness operators had more knowledge of the robot's loca-

*Figure 9.* Team member interaction.

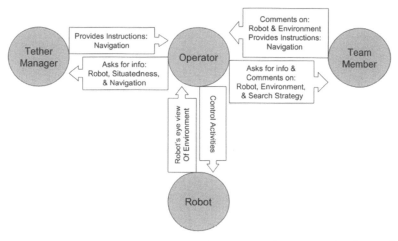

tion and spatial orientation in the void space, and were more focused on goal-directed cues. It follows that the operator's situation awareness is a key factor in successfully planning and executing search operations. Operators with low situation awareness did not seem to have a plan as to how to search using the robot.

Finally, high situation awareness operators engaged in higher levels of reporting (i.e., they talked more to their teammates about situation awareness-related factors in the search environment), $\chi^2(1, N = 2) = 4.74, p < .05$. Talking about the situation helps create a shared mental model of what is happening. Although marginally significant, the data suggest that low situation awareness operators conveyed uncertainty more frequently than did high situation awareness operators, $\chi^2(1, N = 2) = 3.55, p = .06$.

## 6. DISCUSSION

The challenges in perception and situation awareness in human–robot interaction in a USAR environment revealed in our study confirm Sheridan's (1992) findings regarding the difficulties in teleproprioception and telekinesthesis during teleoperation. Our findings suggest one of the main challenges in achieving effective human–robot interaction will be bridging the cognitive gaps between these two entities. The tasks of navigating, searching, mapping, interpreting what is being seen on the video monitor, making decisions about what to do with that information, and dealing with the physical stresses of the job may be overloading the operator in the USAR environment as well. Training and experience may assist the USAR robot operator in cop-

*Figure 10.* Chi-square analyses of statement types comparing operators rated as having high or low situation awareness.

| Variables | Categories | Low SA Operators (Frequency) | High SA Operators (Frequency) | $\chi^2$ | $p$ |
|---|---|---|---|---|---|
| Dyad | Operator–tether manager | 9 | 36 | 16.2 | .000** |
| | Operator–team member | 2 | 24 | 18.6 | .000** |
| | Operator–researcher | 56 | 52 | .15 | .70 |
| | Operator–other | 5 | 4 | .11 | .74 |
| | Operator–group | 42 | 32 | 1.35 | .25 |
| Form | Question | 16 | 37 | 2.81 | .09 |
| | Instruction | 1 | 7 | 4.5 | .03* |
| | Answer | 46 | 49 | .09 | .76 |
| | Comment | 51 | 65 | 1.69 | .19 |
| Topic | State of the robot | 30 | 30 | 0 | 1 |
| | State of the environment | 14 | 20 | 1.06 | .30 |
| | State of Information gathered | 3 | 5 | .5 | .48 |
| | Robot situatedness | 11 | 25 | 5.4 | .02* |
| | Search strategy | 9 | 32 | 12.9 | .000** |
| | Navigation | 32 | 24 | 1.14 | .29 |
| | Off task | 8 | 15 | 2.13 | .14 |
| | Missing | 0 | 4 | 4 | .04* |
| Function | Seek information | 10 | 17 | 1.81 | .17 |
| | Reporting | 22 | 39 | 4.74 | .03* |
| | Clarify | 5 | 6 | .09 | .76 |
| | Confirm | 5 | 11 | 2.25 | .13 |
| | Convey uncertainty | 13 | 5 | 3.55 | .06 |
| | Provide information | 36 | 47 | 1.46 | .23 |
| | Plan | 10 | 16 | 1.38 | .24 |
| | Missing | 13 | 7 | 1.8 | .18 |

*Note.* $N = 2$. SA = situation awareness.
*$p < .05$. **$p < .001$.

ing with stress and the many tasks to be accomplished, and in forming a mental model of how "robot's eye" information is conveyed and then interpreted. We conclude from our data, however, that the information being received from the robot does not match the operator's current mental model. One explanation may be that the perceptual cues (e.g., the keyhole effect noted by Woods et al., 2004) are challenging the operator, and that is where the cognitive deficits begin to appear. This difficulty in integrating the "robo-immersed view" with expectancies regarding the search process mirrors Casper's (2002) observations at the World Trade Center. In both cases, fatigue certainly played a part; it seems likely, however, that lack of a mental model of how a robot "sees" is also a factor.

Videotaped recordings of the robot's eye view during the five operator deployments revealed an almost even split between the amounts of time operators spent actually moving the robot (51%) compared with allowing it to remain stationary (49%.) The percentage of time the robot spent stationary is very similar to the percentage of statements devoted to situation awareness Levels 1 and 2 (both are around 50%). In future work, we plan to study the association of operator statements and robot movements.

Operators discussed search strategy with their teammates using information about the environment, and relating this information to what they already knew. However, only 16% of their statements concerned the state of the environment, or related what they were seeing to known information, a telling percentage in light of the necessity of this information in search operations. The effective use of team processes and communications to compensate for the lack of situation awareness suggests there is an interaction between situation awareness and team communication. Operators with high situation awareness talked to their teammates more about search strategies and robot situatedness; gave more instructions; and reported more on the state of the environment, robot, and information gathered.

Last, quantitative analyses confirm previous research on human–robot interaction in search and rescue operations (Casper, 2002; Casper & Murphy, 2003), which suggested that these tasks will be short and require two operators, not one. Time on task with the robots was of short duration, with the average deployment drop lasting less than 15 min. (Time on task describes the time elapsed from the initial drop of the robot until the conclusion of the operator's run.) Actual drop times at the World Trade Center were even less, averaging 6 to 7 min (Casper, 2002). The ability to complete the search in a short time is a significant factor in the rescue worker's perception of the utility of a rescue robot. As new control tasks evolve utilizing the robots (e.g., carrying medical payloads to victims), operators may spend longer periods of time deploying them.

## 7. CONCLUSIONS AND RECOMMENDATIONS

This study reports on human–robot interaction data from a disaster response training exercise conducted on a collapsed building site. Although the number of operators video recorded is small, the data are rich and the findings lend support to, and add to, prior research. Work in the USAR domain and results from the World Trade Center indicate that perception and communication of perceived aspects of the environment, not navigation, are more significant than previously thought. The major findings of the study lead to the following conclusions and recommendations:

- Cognitive augmentation in the form of intelligent perceptual assistance is needed. On average, the operator is actively engaged in the search task only 32% of the time. In addition, 54% of the operators' statements centered on perception and comprehension of the robot and environment. Finally, the amount of time the robot was stationary was close to 50%. This suggests that it is extremely difficult for operators to establish situation awareness due to inherent perceptual challenges (the world is being perceived from an unnatural viewpoint, the lighting is uncontrolled, etc.) and lack of information in the user interface about the state of the robot (Is it upside down? What pose is it in?). This is consistent with the results of the previous studies of human–robot interaction in USAR (Casper & Murphy, 2002, 2003).
- Robot-assisted technical search is a team task rather than an individual one. The human–robot ratio was never less than 2:1, in part because physical robot operations require, at a minimum, an operator and a tether manager. In addition, the search task itself demands information exchange among team members. More frequent communication with team members was related to higher ratings of operator situation awareness (see Figure 10). Furthermore, operator–team member communication was significantly related to statements involving searches, instructions, and states of information gathered (Figure 7).
- Robot operators need a new cognitive mental model to filter and comprehend data provided by the robot, and to plan effective search strategies. More than one half of operator statements were related to perception and comprehension of the robot and the environment perceived through the robot's eye view. Even so, the low frequency of statements regarding information needed to plan a search (the state of the environment, 13%; state of information gathered, 3%) suggests operators had difficulty reconciling information obtained from the robot's eye view with their existing knowledge of the search environment.
- USAR technical search teams need a new shared mental model of the technical search task to coordinate activities effectively. Operators and their teammates did not have shared expectations regarding their roles in the search process (Figure 8). Operators saw tether managers as a resource for obtaining information about the robot in the environment along with navigation, whereas tether managers saw their role as primarily providing assistance with navigation. Similarly, the operator

saw team members as problem holders, sharing pertinent information about the state of the robot and the environment, and collaborating on search strategy; however, team members did not address operator needs regarding search strategy (Figure 9).

Although the results of this study are preliminary, and must be replicated, the findings give rise to numerous new questions:

- Is the amount of time spent building or maintaining situation awareness stable, or will it change as operators gain experience?
- Will cognitive augmentation shorten the time operators spend gaining situation awareness?
- What perceptual cues are critical for gaining situation awareness in technical search operations?
- Will shared mental models of the robot, the environment, and the search task improve operator performance in search operations?

Future research should examine these and other issues that emerge as USAR personnel acquire more experience working with increasingly sophisticated robotic technology. In particular, the use of visual information (the robot's eye view) as a resource has implications for new ways of conducting USAR operations. Sharing the robot information across various problem holders in the organization (structural and medical specialists, incident commanders) could prove invaluable in reducing the time required to rescue disaster victims. The fact that these problem holders may be physically remote suggests that distributing robot information could reduce the effects of cognitive fatigue or localized noises and distractions that accompany search activities.

In addition, the RASAR CCS generated to organize and examine human–robot interaction may provide insight into the nature of the man–machine relation in USAR and in other robotic domains as robots continue to evolve and become a part of the workplace. Patterns of team process and communication may emerge through analysis that will be useful in training. In the future, robot operators may train in teams rather than individually to capitalize on the interaction between situation awareness and team communication.

---

## NOTES

*Acknowledgments.* We thank Jean Scholtz and Ron Brachman for their support. Many thanks are due to Jenn Casper, Mark Micire, and Brian Minten for their help in collecting the data; Thomas Fincannon for his assistance in editing and transcribing the videotapes; and Rescue Training Associates for providing the test venue. We also thank the editors of this special issue and the reviewers for their detailed, helpful comments on previous versions of this article.

*Support.* This work was supported in part by DARPA under the Synergistic Cyber-Forces Seedling program (N66001–1074–411D) and the Cognitive Systems Exploratory Effort program (N66001-03–8921), and SAIC, Inc.

*Authors' Present Addresses.* Jenny Burke, Graduate Research Assistant, Center for Robot-Assisted Search and Rescue, Department of Computer Science and Engineering, University of South Florida, 4202 E. Fowler Avenue, ENB 342, Tampa, FL 33620. E-mail: **jlburke4@csee.usf.edu**. Robin R. Murphy, University of South Florida, Department of Computer Sciences and Engineering, 4202 E. Fowler Avenue, Tampa, FL 33620. E-mail: **murphy@csee.usf.edu**. Michael D. Coovert, University of South Florida, Department of Psychology, 4202 E. Fowler Avenue, Tampa, FL 33620. E-mail: **coovert@luna.cas.usf.edu**. Dawn Riddle, University of South Florida, Department of Computer Sciences and Engineering, 4202 E. Fowler Avenue, Tampa, FL 33620. E-mail: **riddle@luna.cas.usf.edu**.

*HCI Editorial Record.* First manuscript received December 6, 2002. Revision received July 7, 2003. Accepted by Sara Kiesler and Pamela Hinds. Final manuscript received September 15, 2003. — *Editor*

# REFERENCES

Arkin, R., Fujita, M., Takagi, T., & Hasegawa, R. (2003). An ethological and emotional basis for human–robot interaction. *Robotics and Autonomous Systems, 42,* 191–201.

Breazeal, C. (2000). *Sociable machines: Expressive social exchange between humans and robots.* Unpublished doctoral dissertation, Department of Electrical Engineering and Computer Science, MIT Press, Cambridge, MA.

Breazeal, C. (2003). Toward sociable robots. *Robotics and Autonomous Systems, 42,* 167–175.

Burke, J. L., Murphy, R. R., Rogers, E., Lumelsky, V., & Scholtz, J. (2004). Final report for the DARPA/NSF interdisciplinary study on Human–Robot Interaction. *IEEE Transactions on Systems, Man, and Cybernetics, Part A, 34,* 2.

Casper, J. (2002). *Human–robot interactions during the robot-assisted Urban Search and Rescue response at the World Trade Center.* Unpublished master's thesis, Computer Science and Engineering, University of South Florida, Tampa, FL.

Casper, J., & Murphy, R. (2002). Workflow study on human–robot interaction in USAR. *Proceedings of the 2002 International Conference on Robotics and Automation.* Piscataway, NJ: IEEE.

Casper, J., & Murphy, R. (2003). Human–robot interactions during the robot-assisted search and rescue response at the World Trade Center. *IEEE Transactions on Systems, Man and Cybernetics, Part B, 33,* 367–385.

Cohen, J. A. (1960). Coefficient of agreement for nominal scales. *Educational and Psychological Measurement, 20,* 37–46.

Draper, J., Pin, F., Rowe, J., & Jansen, J. (1999). Next generation munitions handler: Human–machine interface and preliminary performance evaluation. *Proceedings of the 8th International Topical Meeting on Robotics and Remote Systems.* LaGrange Park, IL: American Nuclear Society.

Endsley, M. (1988). *Design and evaluation for situation awareness enhancement. Proceedings of the Human Factors Society 32nd Annual Meeting* (Vol. 1). Santa Monica, CA: Human Factors Society.

Endsley, M. (2000). Theoretical underpinnings of situation awareness: A critical review. In M. Endsley & D. Garland (Eds.), *Situation awareness: Analysis and measurement* (pp. 3–32). Mahwah, NJ: Lawrence Erlbaum Associates, Inc.

Fong, T., Thorpe, C., & Baur, C. (2001). Collaboration, dialogue, and human–robot interaction. *Proceedings of the 10th International Symposium of Robotics Research.* London: Springer-Verlag.

Jones, D., & Endsley, M. (1996). Sources of situation awareness errors in aviation. *Aviation, Space and Environmental Medicine, 67,* 507–512.

Jones, H., & Hinds, P. (2002). Extreme work groups: Using SWAT teams as a model for coordinating distributed robots. *Proceedings of the CSCW 2002 Conference on Computer Supported Cooperative Work.* New York: ACM.

Kawamura, K., Nilas, P., Muguruma, K., Adams, J., & Zhou, C. (2003). An agent-based architecture for an adaptive human–robot interface. *Proceedings of the Hawaii International Conference on System Sciences.* Los Alamitos, CA: IEEE Computer Society.

Khatib, O., Yokoi, K., Brock, O., Chang, K., & Casal, A. (1999). Robots in human environments: Basic autonomous capabilities. *International Journal of Robotics Research, 18,* 684–696.

Kiesler, S., & Goetz, J. (2002). Mental models of robotic assistants. *CHI 2002 extended abstracts.* New York: ACM.

Langle, T., & Worn, H. (2001). Human–robot cooperation using multi-agent systems. *Journal of Intelligent and Robotics Systems, 32,* 143–159.

Murphy, R. (2002). Rats, robots, and rescue. *IEEE Intelligent Systems, 17,* 7–9.

Noldus, L., Trienes, R., Hendriksen, A., Jansen, H., & Jansen, R. (2000). The Observer Video-Pro: New software for the collection, management, and presentation of time-structured data from videotapes and digital media files. *Behavior Research Methods, Instruments and Computers, 32,* 197–206.

Peterson, L., Bailey, L., & Willems, B. (2001). *Controller-to-controller communication and coordination taxonomy (C$^4$T)* (DOT/FAA/AM–01/19). Washington, D.C.: Department of Transportation, Federal Aviation Administration, Office of Aerospace Medicine.

Prince, C., & Salas. E. (2000). Team situation awareness, errors, and crew resource management: Research integration for training guidance. In M. Endsley & D. Garland (Eds.), *Situation awareness analysis and measurement* (pp. 325–347). Mahwah, NJ: Lawrence Erlbaum Associates, Inc.

Sachs, J. (2000). Using a small sample Q sort to identify item groups. *Psychological Reports, 86,* 1287–1294.

Severinson-Eklundh, K., Green, A., & Huttenrauch, H. (2003). Social and collaborative aspects of interaction with a service robot. *Robotics and Autonomous Systems, 42,* 223–234.

Sheridan, T. (1992). *Telerobotics, automation, and human supervisory control.* Cambridge, MA: MIT Press.

Sonnenwald, D., & Pierce, L. (2000). Information behavior in dynamic group work contexts: Interwoven situational awareness, dense social networks and contested collaboration in command and control. *Information Processing and Management, 36,* 461–479.

Stout, R., Cannon-Bowers, J., Salas, E., & Milanovich, D. (1999). Planning, shared mental models, and coordinated performance: An empirical link is established. *Human Factors, 41,* 61–71.

United States Federal Emergency Management Agency. (1992). *Urban search & rescue response plan.* Washington, DC.

Wickens, C. (1992). *Engineering psychology and human performance.* New York: HarperCollins.

Wilkes, D., Alford, A., Cambron, M., Rogers, T., Peters, R., & Kawamura, K. (1999). Designing for human–robot symbiosis. *Industrial Robot, 26,* 47–58.

Woods, D., Tittle, J., Feil, M., & Roesler, A. (2004). Envisioning human–robot coordination for future operation: A roboticist, cognitive engineer and problem holder confront demanding work settings [Special issue]. *IEEE Transactions on Systems, Man, and Cybernetics: Part C, 34,* 1.

HUMAN-COMPUTER INTERACTION, 2004, Volume 19, pp. 117–149

# Beyond Usability Evaluation: Analysis of Human–Robot Interaction at a Major Robotics Competition

**Holly A. Yanco**
*University of Massachusetts Lowell*

**Jill L. Drury**
*The MITRE Corporation*

**Jean Scholtz**
*National Institute of Standards and Technology*

**Holly Yanco** is a roboticist with an interest in navigation in unstructured environments, assistive technology, and methods for improving shared human–robot control; she is an Assistant Professor of Computer Science at the University of Massachusetts Lowell. **Jill Drury** is a usability engineer and researcher with an interest in evaluating collaborative computing systems; she is an Associate Department Head in the Information Technology Center of The MITRE Corporation and an Adjunct Assistant Professor in the Computer Science Department of the University of Massachusetts Lowell. **Jean Scholtz** is a computer scientist with an interest in evaluation of interactive systems and human–robot interaction; she is a research scientist in the Information Access Division of the Information Technology Laboratory at the National Institute of Standards and Technology.

**CONTENTS**

## ABSTRACT

Human–robot interaction (HRI) is a relatively new field of study. To date, most of the effort in robotics has been spent in developing hardware and software that expands the range of robot functionality and autonomy. In contrast, little effort has been spent so far to ensure that the robotic displays and interaction controls are intuitive for humans. This study applied robotics, human–computer interaction (HCI), and computer-supported cooperative work (CSCW) expertise to gain experience with HCI/CSCW evaluation techniques in the robotics domain. As a case study for this article, we analyzed four different robot systems that competed in the 2002 American Association for Artificial Intelligence Robot Rescue Competition. These systems completed urban search and rescue tasks in a controlled environment with predetermined scoring rules that provided objective measures of success. This study analyzed pre-evaluation questionnaires; videotapes of the robots, interfaces, and operators; maps of the robots' paths through the competition arena; post-evaluation debriefings; and critical incidents (e.g., when the robots damaged the test arena). As a result, this study developed guidelines for developing interfaces for HRI.

## 1. INTRODUCTION

When there is a disaster, such as an earthquake or terrorist attack, trained professionals search for victims. Often, these professionals make use of rescue dogs; more recently, they have begun to use robots (e.g., Casper, 2002). Robots will play an even greater role in search and rescue missions in the future because they can squeeze into spaces too small for people to enter and can be sent into areas too structurally unstable or contaminated for safe navigation by human or animal searchers.

Robots have been designed for many situations, including museum guides (Thrun et al., 2000) and conference presenters (Simmons et al., 2003). Urban search and rescue, however, is a prime example of a class of safety-critical situations: situations in which a run-time error or failure could result in death, injury, loss of property, or environmental harm (Leveson, 1986). Safety-critical situations, which are usually also time critical, provide one of the bigger challenges for robot designers due to the vital importance that robots perform exactly as intended and support humans in efficient and error-free operations.

Disasters that can serve as field settings for evaluating robot (and human–robot) performance are rare and unpredictable. Therefore, every year, roboticists hold urban search and rescue competitions to speed the development of research, to learn from one another, and to forge connections to the search and rescue community. The research we describe in this article used one of these competitions to investigate issues in human–robot interaction (HRI). Specifically, we studied HRI at the 2002 American Association for Artificial Intelligence (AAAI) Robot Rescue Competition (also known as AAAI–2002). We focused on the effectiveness of techniques for making human operators aware of pertinent information regarding the robot and its environment.

The study had two parts, centering on the performance of four teams in the AAAI–2002 competition and on the use of two of the teams' systems by a domain expert. The competition provided a unique opportunity to correlate objective performance (e.g, number of victims found, number of penalties assessed, percentage of competition arena area traversed) with user interface (UI) design approaches (e.g., degree of information fusion, presence or absence of a computer-generated map display, etc.). The juxtaposition of the team runs with the domain expert use of interfaces allowed us to compare expert (system developer) versus novice (domain expert) use of the interfaces.

The twin goals of our study were to begin developing a set of HRI design guidelines and, more generally, to gain experience in applying human–computer interaction (HCI) and computer-supported collaborative work (CSCW) techniques to the robotics domain. Although much work has been done in the fields of HCI and CSCW to evaluate the usability of interfaces, little of this work has been applied specifically to robotics.

## 2. RELATED WORK FOR EVALUATION OF HRI

Before any interface (robotic or otherwise) can be evaluated, it is necessary to understand the users' relevant skills and mental models and to develop evaluation criteria with those users in mind. Evaluations based on empirically validated sets of heuristics (Nielsen, 1994) have been used on desktop UIs and Web-based applications. However, current human–robot interfaces differ widely depending on platforms and sensors, and existing guidelines are not adequate to support heuristic evaluations.

Messina, Meystel, and Reeker (2001) proposed some criteria in the intelligent systems literature, but they are qualitative criteria that apply to the performance of the robot only, as opposed to the robots and the users acting as a cooperating system. An example criterion is, "The system ... ought to have the capability to interpret incomplete commands, understand higher level, more abstract commands, and to supplement the given command with additional information that helps to generate more specific plans internally" (p. 1).

In contrast, Scholtz (2002) proposed six evaluation guidelines that can be used as high-level evaluation criteria:

1. Is the necessary information present for the human to be able to determine that an intervention is needed?
2. Is the information presented in an appropriate form?
3. Is the interaction language efficient for both the human and the intelligent system?
4. Are interactions handled efficiently and effectively—both from the user and the system perspective?
5. Does the interaction architecture scale to multiple platforms and interactions?
6. Does the interaction architecture support evolution of platforms?

Usability evaluations use effectiveness, efficiency, and user satisfaction as metrics for evaluation of UIs. Effectiveness metrics evaluate the performance of tasks through the UI. In HRI, the operators' tasks are to monitor the behavior of robots (if the system has some level of autonomy); to intervene when necessary; and to control navigation either by assigning waypoints, issuing a command such as "back-up," or teleoperating the robot if necessary. In addition, in search and rescue, operators have the task of identifying victims and their location.

Not only must the necessary information be present, it must also be presented in such a way as to maximize its utility. Information can be present but in separated areas of the interface, requiring users to manipulate windows to gain an overall picture of system state. Such manipulation takes time and can result in an event not being noticed for some time. Information fusion is another aspect of presentation. Time delays and errors occur when users need to fuse a number of different pieces of information.

As robots become more useful in various applications, we think in terms of using multiple robots. Therefore, the UIs and the interaction architectures must scale to support operators controlling more than one robot.

Robot platforms have made amazing progress in the last decade and will continue to progress. Rather than continually developing new user interaction schemes, is it possible to design interaction architectures and UIs to support hardware evolution? Can new sensors, new types of mobility, and additional levels of autonomy be easily incorporated into an existing UI?

We use Scholtz's (2002) guidelines as an organizing theme for our analysis, operationalizing and tailoring them to be specific to the urban search and rescue environment.

Evaluation methods from the HCI and CSCW worlds can be adapted for use in HRI as long as they take into account the complex, dynamic, and autonomous nature of robots. The HCI community often speaks of three major classes of evaluation methods: inspection methods (evaluation by UI experts), empirical methods (evaluation involving users), and formal methods (evaluation focusing on analytical approaches). Robot competitions lend themselves to empirical evaluation because they involve users performing typical tasks in as realistic an environment as possible (for a description of some robot competitions, see Yanco, 2001). Unfortunately (from the viewpoint of performing the empirical technique known as formal usability testing), robot competitions normally involve the robot developers, not the intended users of the robots, operating the robots during the competition. The performance attained by robot developers, however, can be construed as an "upper bound" for the performance of more typical users. Specifically, if the robot developers have difficulty using aspects of the interface, then typical users will likely experience even more difficulty. In addition, robot competitions afford an interesting opportunity (one not attained so far in formal usability testing of HRI) to correlate HRI performance under controlled conditions to HRI design approaches.

Although the AAAI Robot Competition provided us with an opportunity to observe users performing search and rescue tasks, there were two limitations. First, we were not able to converse with the operators due to the time constraints they were under, which eliminated the possibility of conducting think-aloud (Ericsson & Simon, 1980) or talk-aloud (Ericsson & Simon, 1993) protocols, and also eliminated our ability to have operators perform tasks other than those implied by the competition (i.e., search for victims). Second, the competition simulated a rescue environment. Many of the hazards (beyond those incorporated in the arena) and stress-inducing aspects of an actual search and rescue environment were missing. Nonetheless, this environment was probably the closest we could use in studying search and rescue tasks due to safety and time constraints in actual search and rescue missions.

Two patterns were observed in previous HRI empirical testing efforts that limit the insights obtained to date. The first, as mentioned previously, is a tendency for robot performance to be evaluated using atypical users. For example, Yanco (2000) used a version of a usability test as part of an evaluation of a robotic wheelchair system but did not involve the intended users operating the wheelchair (the wheelchair was observed operating with able-bodied occupants). We have started to break this pattern by also analyzing the use of two urban search and rescue robot systems by a fire chief, a more typical user, after the competition runs were completed.

The second pattern that limits HRI empirical testing effectiveness is the tendency to conduct such tests very informally. For example, Draper, Pin, Rowe, and Jansen (1999) tested the Next Generation Munitions Handler/Advanced Technology Demonstrator, which involves a robot that re-arms military tactical fighters. Although experienced munitions loaders were used as test participants, testing sessions were actually hybrid testing and training sessions, and test parameters were not held constant during the course of the experiment. Data analysis was primarily confined to noting test participants' comments such as, "I liked it when I got used to it." Our study took advantage of the structure inherent in the conduct of the AAAI Robot Competition to keep constant variables such as environment, tasks, and time allowed to complete tasks. In addition, the competition is held annually, which will allow us to track HRI progress and problems over time.

## 3. METHOD

The two portions of the study consisted of evaluating the interfaces as their developers competed and when the domain expert performed four tasks with each of the interfaces. This section begins with descriptions of the criteria we used for evaluating the interfaces and the evaluation environment we used for both portions of the study. It continues with the methodology used for assessing the interfaces as they were used during the competition. We correlated competition performance with various features in the interface design; therefore, we describe the competition scoring methodology in the fourth subsection. The methodology we used for the domain expert runs comprises the fifth subsection. Finally, we coded the resulting videotapes of both portions of the study using the same coding scheme, which we describe at the end of this section.

### 3.1. Method for Assessing Interaction Design

An accepted evaluation methodology in HCI is to take a general set of principles and tailor them for use in evaluating a specific application (e.g., see Nielsen, 1993). We operationalized and tailored Scholtz's (2002) evaluation guidelines as follows to be more specific to the case of HRI in an urban search and rescue context.

"Is the necessary information present for the human to be able to determine that an intervention is needed?" becomes "Is sufficient status and robot location information available so that the operator knows the robot is operating correctly and avoiding obstacles?" "Necessary information" is very broad. In the case of urban search and rescue robots, operators need information regarding the robot's health, especially if it is not operating correctly. Another critical piece of information operators need is the robot's location relative to obstacles, regardless of whether the robot is operating in an autonomous or teleoperated mode. In either case, if the robot is not operating correctly or is about to collide with an obstacle, the operator will need to take corrective action.

"Is the information presented in an appropriate form?," becomes "Is the information coming from the robots presented in a manner that minimizes operator memory load, including the amount of information fusion that needs to be performed in the operators' heads?" Robotic systems are very complex. If pieces of information that are normally considered in tandem (e.g., video images and laser ranging sensor information) are presented in different parts of the interface, the operator will need to switch his attention back and forth, remembering what he or she saw in a previous window to fuse the information mentally. Operators can be assisted by information presentation that minimizes memory load and maximizes information fusion.

"Is the interaction language efficient for both the human and the intelligent system? Are interactions handled efficiently and effectively—both from the user and the system perspective?" Combining these two, they become, "Are the means of interaction provided by the interface efficient and effective for the human and the robot (e.g., are shortcuts provided for the human)?" We consider these two guidelines together because there is little language per se in these interfaces; rather, the more important question is whether the interactions minimize the operator's workload and result in the intended effects.

We are looking at interaction in a local sense, that is, we are focused on interactions between an operator and one or more robots. The competitions currently emphasize this type of interaction but do not provide an environment to study the operator–robot interaction within a larger search and rescue team.

Interactions differ depending on autonomous capabilities of the robots. From the user perspective, we are interested in finding the most efficient means of communicating with robots at all levels of autonomy. For example, if a robot is capable of autonomous movement between waypoints, then how does the operator specify these points? The interaction language must also be efficient from the robot point of view. Can the input from the user be quickly and unambiguously parsed? If the operator inputs waypoints by pointing on a map, what is the granularity? If the user types robot commands, is the syntax of the commands easily understood? Are error dialogues needed in the case of missing or erroneous parameters?

"Does the interaction architecture scale to multiple platforms and interactions?" becomes "Does the interface support the operator directing the actions of more than one robot simultaneously?" A goal in the robotics community is for a single operator to be able to direct the activities of more than one robot at a time. Multiple robots can allow more area to be covered, can allow for different types of sensing and mobility, or can allow for the team to continue operating after an individual robot has failed. Obviously, if multiple robots are to be used, the interface needs to enable the operator to switch his or her attention among robots successfully.

"Does the interaction architecture support evolution of platforms?" becomes "Will the interface design allow for adding more sensors and more autonomy?" A robotic system that currently includes a small number of sensors is likely to add more sensors as they become available. In addition, robots will become more autonomous, and the interaction architecture will need to support this type of interaction. If the interaction architecture has not been designed with these possibilities in mind, it may not support growth.

## 3.2. Assessment Environment

The robots competed in the Reference Test Arenas for Autonomous Mobile Robots developed by the National Institute of Standards and Technology (NIST; Jacoff, Messina, & Evans, 2000, 2001). The arena consists of three sections that vary in difficulty. The yellow section, the easiest to traverse, is similar to an office environment containing light debris (fallen blinds, overturned table and chairs). The orange section is more difficult to traverse due to the variable floorings, a second story accessible by stairs or a ramp, and holes in the second story flooring. The red section, the most difficult section, is an unstructured environment containing a simulated pancake building collapse, piles of debris, unstable platforms to simulate a secondary collapse, and other hazardous junk such as rebar and wire cages. Figure 1 shows one possible floor plan for the NIST arena. The walls of the arena are easily modified to create new internal floor layouts, which prevent operators from having prior knowledge of the arena map.

In the arena, victims are simulated using mannequins. Some of the mannequins are equipped with heating pads to show body warmth, motors to create movement in the fingers and arms, tape recorders to play recordings of people calling for help, or all three. Victims in the yellow arena are easier to locate than victims in the orange and red arenas. In the yellow arena, most victims are located in the open. In the orange arena, victims are usually hidden behind obstacles or on the second level of the arena. In the red arena, victims are in the pancake layers of the simulated collapse. Between rounds, the victim locations are changed to prevent knowledge gained during earlier rounds from providing an easier search in later rounds.

Operator stations were placed away from the arena and set up so that the operators would have their backs to the arena. Therefore, the operators were not able to see the progress of their robots in the arena; they had to assess the robots' situations using their UIs.

## 3.3. Method for Studying Team Performance

Teams voluntarily registered for the competition. We asked them to participate in our study but made it clear that study participation was not a requirement for competition participation. The incentive to participate in the study was the chance to have their robot system used by a domain expert in the second part of the study.

*Figure 1.* The National Institute of Standards and Technology test arena for urban search and rescue used by the robots in the competition.

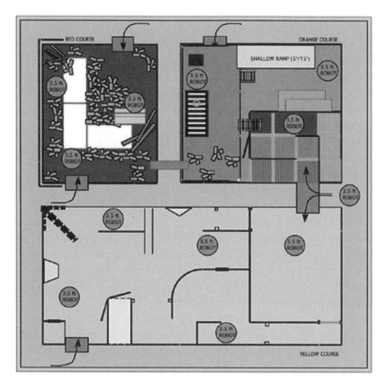

Participating teams were asked to fill out a questionnaire before the start of the competition. The questions inquired about the robot hardware being used, the type of data provided to the human operator, the level of autonomy achieved by the robot, the maturity of the robot design, and whether the interface was based on a custom (bespoke) or commercial product.

Once the competition began, we observed the operator of each team's robots during the three 15-min runs of the competition. The operator and the interface screen were videotaped. The robots were also videotaped in the arena; cameras were placed in various locations around the arena in an attempt to keep the robot constantly within sight.

We were silent observers, not asking the operators to do anything differently during the competition than they would have already done; our study could not impact on the competition outcome. We could not ask the participants to use the "thinking aloud" protocol, although one participant who was eager to obtain feedback on his interface voluntarily voiced his thoughts as he worked with his robot during the competition. At the conclusion of each run, our observer performed a quick debriefing of the operator via a short postrun interview to obtain the operator's assessment of the robot's performance.

In addition, we were given the scoring materials from the competition judges that indicated where victims were found and penalties that were assessed. We also created maps by hand that showed the approximate paths that the robots took and marked the critical incidents such as hitting objects or victims that occurred during the runs.

## 3.4. Method for Scoring Team Performance

The scoring algorithm utilized the number of victims found and the accuracy of reporting the location of the victims. The scoring scheme penalized teams for allowing robots to bump into obstacles or victims.[1] The judges recorded a minor victim penalty for bumping into a victim (subtracting .25 from the number of victims found), and a major victim penalty was scored for an event such as causing a pancake layer to collapse on a victim (subtracting 1). Minor damage to the environment, such as moving a wall a small amount, was marked as a minor environment penalty (subtracting .25), whereas major environment damage, such as causing a secondary collapse, was considered a major penalty (subtracting .75).

The scoring formula is:

$$Performance\ Score = (V - P) * A$$

where $V =$ number of victims found; $P =$ penalties; and $A =$ accuracy $= 1$ if map produced by system, 0.6 if good quality hand-drawn map produced, 0.4 if poor hand-drawn map produced (the accuracy score was determined by the competition judges).

## 3.5. Method for Studying Domain Expert Performance

After the competition, we had access to a search and rescue domain expert: a special operations fire chief who had participated in training sessions with robots for search and rescue. The goals of the evaluation were to assess ease of learning as well as ease of use. To evaluate ease of learning, the domain expert was asked to explore the interface for 5 min to determine what information was available in the interface. Then the domain expert was given about 5 min

---

1. The scoring algorithm used for comparing teams in this study differs from the official scoring algorithm used in the competition (AAAI/RoboCup, 2002). We factored out measures that were unrelated to the interface, such as a measure for calculating a bonus when unique victims were found by different robots.

of training, which would be a realistic amount of training in the field in an emergency condition if the primary (more thoroughly trained) operator suddenly became unavailable (R. Murphy, personal communication, August 2002). After the training, the domain expert was asked to describe the information available in the interface that he did not see during the initial exploration period. Finally, the domain expert was asked to navigate the robot through the arena.

The domain expert was able to verbalize his thoughts as he navigated the robots. He produced a combination of think-aloud and talk-aloud protocols. In general, as he was navigating through the arena, he used the talk-aloud protocol. However, there were a number of times when we experienced technical difficulties, and the chief had to wait for a resolution to the problem before he could proceed. During these times, his verbalizations were more introspective.

## 3.6. Method for Coding Team and Domain Expert Sessions

Our data consisted of videotapes, competition scoring sheets, maps of robot paths, questionnaire and debriefing information, and researcher observation notes. The richest sources of information were the videotapes. In most cases, we had videotapes of the robots moving through the arena, the UIs, and videos of the operators themselves.

To make the most of the videotaped information, we developed a coding scheme to capture the number and duration of occurrences of various types of activities observed. Our scheme consists of a two-level hierarchy of codes: Header codes capture the high-level events, and primitive codes capture low-level activities. The following header codes were defined: identifying a victim, robot logistics (e.g., undocking smaller robots from a larger robot), failures (hardware, software, or communications), and navigation and monitoring navigation (directing the robot or observing its autonomous motion). Three primitive codes were defined: monitoring (watching the robot when it is in an autonomous mode), teleoperation ("driving" the robot), and UI manipulation (switching among windows, selecting menu items, working with dialog boxes, typing commands, etc.).

Our coding scheme was inspired by the structure of the Natural Goals, Operators, Methods, and Selection Rules Language (NGOMSL) used to model UI interaction (Kieras, 1988). NGOMSL models consist of a top-down, breadth-first expansion of the user's top-level goals into methods, and the methods contain only primitive operations (operators), branch statements, and calls to other NGOMSL methods. Our top-level header codes can be thought of as NGOMSL goal-oriented methods for identifying a victim, navigation and monitoring, or handling robot logistics or failures. Our primitives are not always true primitives (e.g., an activity such as teleoperation can usually be broken down into finer grained motor control actions). However, they are at the lowest level that it makes sense to analyze, and thus they are analogous to NGOMSL primitives.

The coding was done by two sets of researchers. To obtain intercoder reliability, both sets initially coded the same run and compared results. The kappa computed for agreement was .72 after chance was excluded.[2] We then discussed the disagreements and, based on a better understanding, we coded the remaining runs. We did not formally check interrater reliability on the remaining runs as we found in the initial check that we easily agreed on the coding for the events that were observable, but noted that the timing of those events could only be determined within a few seconds. Unfortunately, we could not see the robot when it was in a covered area or when it was in the small portions of the arena that the cameras did not cover.

## 4. DESCRIPTIONS OF SYSTEMS STUDIED

Eight teams entered the competition. However, we only investigated the HRI of the four teams who found victims during their runs; these teams were also the top-ranked teams. Teams that were unable to find victims most often had hardware failures and no significant amount of HRI to study. In this section, we describe each of the four systems in our study, including the UI and the robot hardware. A summary of the systems is given in Figure 2.

### 4.1. Team A

Team A developed a heterogeneous team of five robots—one iRobot® ATRV-Mini and four Sony AIBOs®, for the primary purpose of research in computer vision and multiagent systems.[3] They spent 3 months developing their system for the rescue competition. All robots were teleoperated serially. The AIBOs were mounted on a rack at the back of the ATRV-Mini. The AIBOs needed to be undocked to start their usage and redocked after they were used if the operator wanted to continue to take them with the larger robot.

Team A developed two custom UIs, which were created for use by the developers and were not tested with other users before the competition. There was one UI for the ATRV-Mini and another for the AIBOs. The UIs ran on separate computers. Communication between the UI and the robots was accomplished using a wireless modem (802.11b).

---

2. When chance was not factored out, the agreement was .8.

3. The identification of any commercial product or trade name does not imply endorsement or recommendation.

*Figure 2.* **Summary of system characteristics.**

| Variables | Team A | Team B | Team C | Team D |
|---|---|---|---|---|
| Platform and autonomy characteristics | One iRobot ATRV-Mini and four Sony AIBOs, teleoperated serially | One iRobot ATRV-Jr. with a range of operator-selectable autonomy levels | Two RWI Magellan Pros, teleoperated or avoid obstacles while moving toward goals | Two custom robots (one wheeled and one tracked), teleoperated serially |
| Sensors | ATRV-Mini: Video (circle of 8 cameras), current laser scan (raw and processed for map); odometry and laser scan fused for map. AIBOs: Video, no fusion | Video, thermal imaging (raw), infrared, bump, laser scan, and sonar; data from last four sensors fused for sensor map | Video, sonar, infrared; data from last two sensors fused for overhead map (evidence grid) | Both have video and no sensor fusion |
| Interfaces | ATRV-Mini: Multiple windows for video, map, raw laser scan, camera control; keyboard control. AIBOs: video window; keyboard or GUI control | Touch screen with windows for sensor status, battery/velocity/tilt, video (actually displayed on another monitor), sensor map, environment map; control via joystick and touch screen | GUI: Split screen for two robots with video on top and map on bottom. Text-based interface: 14 text and four graphic windows. Control via keyboard | Two displays used: One for video feed, other for pre-entered map of arena; control via keyboard |

*Note.* GUI = graphical user interface.

*Figure 3.* (a) **Team A's interface for the iRobot ATRV-Mini.** (b) **Team A's interface for the Sony AIBOs.**

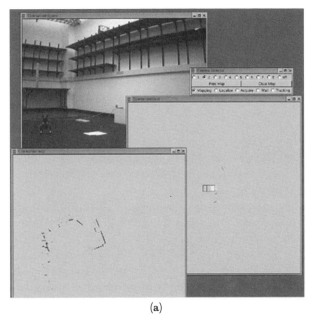

(a)

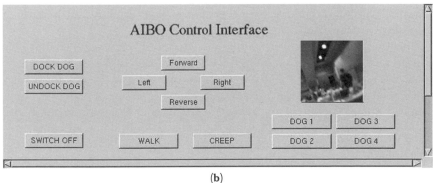

(b)

The UI for the ATRV-Mini, shown in Figure 3a, had multiple windows. In the upper left-hand corner was a video image taken by the robot, updated once or twice each second. In the lower left-hand corner was a map constructed by the robot using the SICK® laser scanner and odometry (SICK, Inc., Bloomington, MN). In the lower right-hand corner, the raw laser scan information was presented as lines showing distance from the robot. The upper right-hand corner had a window with eight radio buttons labeled 1 through 8 to allow the user to switch camera views. The operator drove the robot using keys on the keyboard to move forward, backward, right, and left.

The UI for the AIBOs, shown in Figure 3b, had a window with the video image sent from the robot. The operator controlled the robots either using buttons on the graphical UI (GUI) or by using the keyboard. (The domain expert controlled the AIBOs using the keyboard because of a problem with the GUI at that time.)

## 4.2. Team B

Team B had been developing their robot system for use in hazardous environments for less than 1 year. The robot was an iRobot ATRV-Jr. Communication was achieved through a proprietary, low-bandwidth communication protocol over 900 MHz radio.

The custom UI, shown in Figure 4, was developed for expert users and tested with novice users and experienced operators. The interface was displayed on a touch screen. The upper left-hand corner of the interface contained the video feed from the robot. Tapping the sides of the window moved the camera left, right, up, or down. Tapping the center of the window recentered the camera. (During the competition, the video window had not yet been finished, so the video was displayed on a separate monitor. However, the blank window was still tapped to move the camera.) The robot was equipped with two types of cameras that the operator could switch between: a color video camera and a thermal camera.

The lower left-hand corner contained a window displaying sensor information such as battery level, heading, and tilt of the robot. In the lower right-hand corner, a sensor map was displayed, showing filled red areas to indicate blocked directions. In the picture of the previous interface, a map of the environment can be seen in the upper right-hand corner. Although this space was left for a map during the competition, the software for building and displaying maps had not yet been created, so no maps were provided to the operator.

The robot was controlled through a combination of a joystick and the touch screen. To the right of the sensor map, there were six mode buttons: Auto (autonomous mode), Shared (shared mode, a semi-autonomous mode in which the operator can "guide" the robot in a direction but the robot does the navigation and obstacle avoidance), Safe (safe mode, in which the user controls the navigation of the robot, but the robot uses its sensors to prevent the user from driving into obstacles), Tele (teleoperation mode, in which the human controller is totally responsible for directing the robot), Escape (a mode not used in the competition), and Pursuit (also not used in the competition). Typically, the operator would click on one of the four mode buttons then start to use the joystick to drive the robot. When the operator wished to take a closer look at

*Figure 4.* **Team B's user interface.**

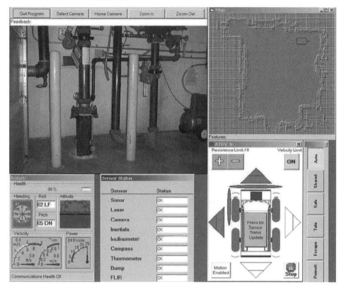

something, perhaps a victim or an obstacle, he would stop driving and click on the video window to pan the camera. For victim identification, the operator would switch over to the thermal camera for verification.

## 4.3. Team C

Team C had developed their robots for less than 2 years as a research platform for vision algorithms and robot architectures. They used two identical robots, RWI® Magellan Pros. Communication between the UI and robots was achieved with an radio frequency modem.

The robots had a mixed level of autonomy: They could be fully teleoperated, or the robots could provide obstacle avoidance while achieving a specified goal. The robots could run simultaneously, but were operated serially. Waypoints were used to generate maps from the robot's current location to the starting point. The operator commanded the robot by giving it relative coordinates to move toward. The robot then autonomously moved to that location using reactive obstacle avoidance. The robot's ability to carry out a command without assistance allowed for the perception that the operator is moving both robots "at once," although he was controlling them serially. It was the operator's trust in the robots' autonomy that allowed this type of operation; the operator did not need to monitor the progress of one robot while commanding the other.

*Figure 5.* **Team C's graphical user interface.**

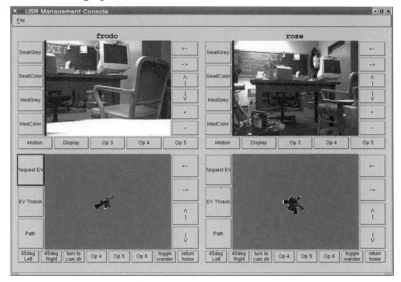

A custom interface, shown in Figure 5, was developed for a "sophisticated user" (according to the developers). Team C started Run 1 using a GUI, but switched to a text-based interface when there were command latency problems with the GUI. In the GUI, the screen was split down the middle; each side was an interface to one of the two robots. The top window for each robot displayed a current video image from the robot, and the bottom window displayed map information.

In the alternative text-based interface, used in the remainder of the runs, the screen had 14 text windows and four graphic windows, one half for each of the robots. Seven text windows were used for the following: the interprocess communication server, the navigation module, the vision module, the mapping module, the navigation command line, a window for starting and monitoring the visual display, and a window for starting and monitoring the map display. Two graphic windows were used for displaying the camera image and the map image. The computer was running an enlarged desktop during the competition, and the operator sometimes needed to switch to another part of the desktop (effectively switching to another display) for other pieces of the interface. The robots were controlled with keystrokes.

## 4.4. Team D

Unlike the other three systems, Team D developed their robots for search and rescue over the previous year. They had custom-built robots, one wheeled and one tracked, both with the same sensing and operating capabilities. The robots were teleoperated serially. A wireless modem was used to communicate between the UI and the robots.

Team 4 developed a custom UI on two screens.[4] One monitor displayed the video feed from the robot that was currently being operated. The other monitor had a pre-entered map of the arena on which the operator would place marks to represent the locations of victims that were found. The robots were driven with keyboard controls.

## 5. RESULTS

We present two types of results for the teams: the objective measures from the competition and the results of our coding. We also present the coded results of the domain expert's performance along with his talk-aloud and think-aloud protocols. Finally, we analyze performance using the Scholtz (2002) guidelines from Section 3.1.

### 5.1. Team Runs

Each team had three 15-min runs during the competition. We only coded Runs 1 and 3 due to the failure of the video data capture equipment during Run 2. The total times are in some cases less or more than the allotted 15 min. It was sometimes difficult to discern the actual starting time for the competition to coordinate the start of data capture, which resulted in shorter times. In addition, in Team A's first run, a tape change caused us to lose some of the data from the run. Longer times resulted from a judge failing to stop the run at exactly 15 min.

Figure 6 shows the percentage of time spent in each of the primary header code activities. The majority of time for most runs was spent navigating, followed by identifying victims. Time spent in logistics or failures was time taken away from looking for victims.

Team scores are shown in Figure 7. Because we did not analyze the HRI in Run 2, we only consider Runs 1 and 3 in the scoring. Using the scoring algorithm in Section 3.4, the rankings for the two rounds would be as follows: 1st place, Team A; 2nd place, Team C; 3rd place, Team D; and 4th place, Team B.[5]

---

4. We were unable to obtain a screen shot of this interface from its designers.

5. The actual rankings in the competition, which included Run 2 as well as other measures such as in which part of the arena victims were found, were as follows: 1st place, Team D; 2nd place, Team C; 3rd place, Team A; and 4th place, Team B.

*Figure 6.* **Time spent in each of the primary header codes for competition runs.**

| Teams | Run | Total Time | % Time Navigation–Monitoring Navigation | Victim ID | Failure | Logistics |
|---|---|---|---|---|---|---|
| A | 1 | 10:39 | 46 | 51[a] | 0 | 3 |
|   | 3 | 14:45 | 62 | 18 | 19[b] | 1[c] |
| B | 1 | 14:33 | 81[d] | 19 | 0 | 0 |
|   | 3 | 16:42 | 77 | 23 | 0 | 0 |
| C | 1 | 13:26 | 59 | 23 | 17[e] | 0 |
|   | 3 | 14:39 | 69 | 12 | 18[f] | 0 |
| D | 1 | 15:12 | 55 | 32 | 0 | 12[g] |
|   | 3 | 13:30 | 87 | 4 | 0 | 9 |

*Note.* ID = identification.
[a] Includes navigation to get a new angle for victim ID after a judge said that the first image was unclear. [b] Wireless modem failures. [c] In addition, about 25% of the victim ID time was spent in logistics while deploying AIBOs. [d] The operator spent 90% of this navigation time in a confused state. However, the equipment had not malfunctioned, so this was not coded as a failure. [e] Graphical user interface latency, panoramic image failure. [f] Panoramic image failure, vision system on one robot failed midway through run. [g] Switching between two robots.

*Figure 7.* **Team scores computed using the algorithm in Section 3.4.**

| Teams | Run | No. of Victims | Penalties | Accuracy | Score | Team Total |
|---|---|---|---|---|---|---|
| A | 1 | 4 | 8 × 0.25 | 1.0 | 2.0 | |
|   | 3 | 3 | 6 × 0.25 | 1.0 | 1.5 | 3.5 |
| B | 1 | 3 | 5 × 0.25 | 0.6 | 1.05 | |
|   | 3 | 0 | 1 × 0.25 + 3 × 0.75 | — | Negative | < 1.05 |
| C | 1 | 3 | 0 | 0.4 | 1.2 | |
|   | 3 | 4 | 3 × 0.25 | 0.6 | 1.95 | 3.15 |
| D | 1 | 6 | 9 × 0.25 | 0.5[a] | 1.875 | |
|   | 3 | 3 | 4 × 0.25 | 0.6 | 1.2 | 3.075 |

[a] This number reflects a penalized accuracy score, as determined by the judges. There was some question as to whether advance knowledge of the arena layout had been obtained.

## 5.2. Domain Expert Runs

### Ease of Learning

The domain expert, a special operations fire chief trained in search and rescue and with experience using robots, used two systems: Team A (teleoperated) and Team B (different autonomy modes). We started each session with a short amount of time for the chief to explore the interface without instruction. After this period, we asked him to state what he could figure out about the interface. Then the system developers explained the interface to him, and we asked the chief what was in the interface that he had not seen before.

For Team B, the chief said that there was no real-time video (the team was having trouble with their video link at this time, so they were only sending about one frame per second). He noted that there were sensors around the robot, pointing to the sensor map in the lower right-hand corner, and that the map appeared to be displaying proximity information. After the chief talked with Team B's developer, he stated that he had learned about the control modes for the robot.

For Team A, the chief said that he saw a laser map on the lower right, a video display on the upper right, an ultrasonic map on the left, and a data window under that. He could not see how to drive, but thought he would do it using the arrow and the mouse. After the developer's explanation, the chief learned that the window on the left did not have an ultrasonic map, but was instead displaying a map created as he drove using the laser scan and odometry. He also learned how to control the robot and that there was a ring of cameras on top of the robot for the video window. A window with radio buttons labeled 1 through 8 was used to switch from one camera view to another in the video window.

**Ease of Use (Performance)**

The chief had been a judge for the competition, so he was more familiar with the arena at the time of his runs than any of the competitors had been. Figure 8 shows the amount of time the domain expert spent in each of the primary header codes. The times shown in the table include the time that the expert was using the systems, not any time that he was speaking to the system developer or the researchers.

We observed the chief relying heavily on the live video for navigation. He would drive, change camera angles, then resume driving. We discuss the primary use of video further in Section 5.3.

## 5.3. Evaluation Using Tailored Scholtz Guidelines

We use the performance of the teams and of the domain expert, the results of coding activities of the operators during the competition runs, and an examination of critical incidents to discuss Scholtz's (2002) guidelines from Section 3.1.

*Figure 8.* **Percentages of run time spent in each of the primary header modes for the domain expert's runs.**

| Teams | Total Time | Navigation–Monitoring Navigation | Victim ID | Failure | Logistics |
|---|---|---|---|---|---|
| | | | % Time | | |
| A | 18:43 | 93 | 0[a] | 2[b] | 5 |
| B | 25:35 | 97 | 3 | 0 | 0 |

Note. ID = identification.
[a] Victims were removed from the arena during the chief's runs. [b] Communication failure: video signal was not updating.

### Is Sufficient Status and Robot Location Information Available so That the Operator Knows the Robot Is Operating Correctly and Avoiding Obstacles?

The number of penalties for each team is shown in Figure 9. Note that Team A's two different types of robots are listed separately, because the ATRV-Mini has dramatically different sensor capabilities than the AIBOs. Although another team, Team D, also fielded robots of different types, their robots differed only in their navigation properties (one type was tracked and the other was wheeled); otherwise, their sensor suites and operational capabilities were identical.

We had thought that Team B would fare slightly better than they did. Team B's operator experienced serious confusion when he forgot that his robot's video camera was pointing in a direction other than straight ahead. This confusion resulted in more than one half of Team B's first run (8½ min) being wasted. The interface did not provide any reminders that the video camera was pointing off center, so this lack of awareness of robot state (rather than a paucity of sensor data) caused him to run into more obstacles and find fewer victims than he might have otherwise.[6] We are unsure why he also had a poor Run 3. During this run, the operator was frustrated that his robot was "too big" to navigate in the small areas of the arena. In fact, he did have the largest robot in the competition.

We saw several specific instances where operators were unaware of robot locations and surroundings. In several cases (e.g., Team D during Run 1), there was not enough awareness of the area immediately behind the robot, causing the robot to bump obstacles when backing up. Even when

---

6. This problem was corrected by the developers before other runs by changing the program to recenter the camera.

*Figure 9.* **Number of penalties incurred by the teams.**

| Teams | Run | Arena Penalties | Victim Penalties | Rank[a] |
|---|---|---|---|---|
| A (ATRV-Mini) | 1 | 1 minor[b] | 0 | |
| | 3 | 4 minor | 0 | 2 |
| A (AIBOs) | 1 | 7 minor[c] | 0 | |
| | 3 | 2 minor | 0 | — |
| B | 1 | 3 minor | 2 minor | |
| | 3 | 1 minor, 3 major | 0 | 3 |
| C | 1 | 0 | 0 | |
| | 3 | 3 minor | 0 | 1 |
| D | 1 | 9 minor | 0 | |
| | 3 | 4 minor | 0 | 4 |

[a] 1 is the best, 4 is the worst bumping record overall based on numbers of bumps. AIBOs are not ranked because they were used for only short periods of time. [b] The ATRV-Mini was used for approximately 12 min during each of Runs 1 and 3. Normalizing to 15 min would result in 1.25 and 5 minor arena penalties, which does not affect Team A's overall ranking. [c] The AIBOs were used for approximately 3 min during Runs 1 and 3.

moving forward, several operators (e.g., Team B during Run 3) hit walls and were not aware of doing so. One of Team A's robots was trapped under fallen Plexiglas®, but the operator was never aware of this situation. Because they did not have precise awareness of the area immediately around the robot, operators (e.g., Team B during Run 3) had a difficult time maneuvering the robots in tight spaces.

One of the debriefing questions we asked after each run was how the operator perceived the performance of the run. Surprisingly, Team B's operator stated after Run 3 that he had not hit anything during the run. However, his perceptions did not correspond with reality; he had incurred one minor and three major arena penalties during this run. Clearly, the operator did not have sufficient awareness of the robot, its surroundings, and its activities.

The chief's bumping performance is shown in Figure 10. Although he was not scored, we marked the times that he hit objects just as was done for the teams. These penalties were marked over the full length of the chief's runs, which were about 10 min longer than an average team run.

While using Team A's system, the chief asked twice if someone was watching in the arena. The first time he said he was not sure if the robot was clear of a wall. The second time he thought the robot might be caught on a cable, but he was told that the robot was clear. To resolve his awareness problems, he deployed an AIBO from the ATRV-Mini and positioned the camera on the ATRV-Mini so that he could view the AIBO while he was teleoperating it.

The chief had begun to experiment with Team B's system earlier and stopped due to wireless interference. He did not feel comfortable relying on

*Figure 10.* Penalties incurred by the chief during his runs.

| Teams | Arena Penalties | Victim Penalties |
|---|---|---|
| A (ATRV Jr.) | 0 | 0 |
| A (AIBOs) | 0 | 0 |
| B | 2 minor, 6 major | 0 |

the sensor display, the single frame video images updated infrequently, and various modes of autonomy for navigation. In this early run, the chief was using the safe mode of navigation and was unable to understand why he could not navigate through a perceived opening. He put the robot in teleoperation and discovered that the "opening" was covered with Plexiglas, but only when people called from the arena area to state that the robot had charged through the panel.

When using both systems, the chief adjusted the camera views frequently, but even then he had difficulty knowing where the robot was. The team operators using these systems relied far less on moving the cameras around to acquire awareness than the chief did. Team B's operator relied on the sensor data and used various modes of autonomy. He used the camera views when he was identifying a victim. However, he also had imperfect awareness; there were a number of instances when he bumped into obstacles and was penalized in the scoring but never noticed this during the run. Team A's operator used the dynamically created map and the laser scanning data for navigating, but he also had suboptimal awareness. When one of the AIBOs fell off the ATRV-Mini, the operator was completely unaware of it.

The developers seemed to feel more comfortable relying on sensor data other than video, which may have provided a false sense of security as their penalty scores reflected their lack of awareness. The chief, a novice user, was more cautious and, although he commented about the usefulness of the sensor data, he still relied heavily on live video feeds that proved to be problematic. Further, not all of the necessary information was presented to users; more information was needed regarding the awareness of the relation of the robot to its environment, as evidenced by a number of bumping incidents.

## Is the Information Coming From the Robots Presented in a Manner That Minimizes Operator Memory Load, Including the Amount of Information Fusion That Needs To Be Performed in the Operators' Heads?

Team D had the only system in the competition that had no information fusion in the system, using only video. Team A had the only system that presented a map in the display that included the walls of the arena. This map allowed the operator to see where he had been so that he could hopefully avoid covering the same territory numerous times. Team C also had a map in their interface, but it presented only the sonar readings of the robot as it moved through the arena. No corrections were made for dead reckoning errors. Although Team A's map looked like a floor plan, Team C's map looked like a fat line composed of black triangles. Figure 11 shows that Team A had better coverage than all teams, with the exception of Team D for Run 1.[7] The two teams with maps, A and C, scored above (1st and 2nd, respectively) the two teams without maps, B and D (4th and 3rd, respectively); see Figure 7 for a summary of the scoring.

Although additional sensor information should provide additional awareness as a general rule, this rule does not hold true if more information is provided but the information is not integrated into the displays in a way that an operator can use. In general, lack of data fusion, other than that contained in maps, hindered operators' ability to quickly obtain an understanding of the robot's status and location. For example, for Teams A and B, the video image was presented separately from the sonar or laser ranging sensor data, in opposite corners of the display screen. Such separation requires the operator to mentally synthesize the data as opposed to having the interface provide a combined picture.

Presenting related data in opposite corners of the display is an example of how the displays were not laid out for maximum efficiency or memory load minimization. Evidence of this trend can be seen by the fact that operators spent a large percentage of the time in UI manipulation. Various types of information were, in general, presented in separate windows so that operators spent significant time periods moving between windows. Operators then had to remember what was in one window and combine it with information in other windows. Some operators needed to constantly glance between video and other data, or move between windows on the display, while mentally fusing the various pieces of information.

When using System A, the chief initially noted that he relied primarily on the laser for navigation. However, his primary navigation method was to stop teleoperating the robot and to change the view of the camera to look around. The chief used a similar method to drive Team B's system. He relied heavily on live video and commented when the reception was particularly bad. However, video can miss some types of obstacles, as evidenced by the fact that the chief drove through a Plexiglas panel.

During the first run, the Team B operator moved the robot's video camera off center to look at a victim for identification, and also switched to his thermal

---

7. There was some question as to whether Team D had prior knowledge of the arena for Run 1.

*Figure 11.* **Amount of the arena covered.**

| Teams | Run | Coverage |
|-------|-----|----------|
| A | 1 | 50% yellow |
|   | 3 | 35% yellow |
| B | 1 | 20% yellow, 5% orange |
|   | 3 | 35% yellow, 10% orange |
| C | 1 | 30% yellow |
|   | 3 | 35% yellow |
| D | 1 | 80% yellow |
|   | 3 | 15% yellow, 10% orange, 5% red |

camera to verify that it was a live victim. After the victim identification, the operator switched to shared mode to allow the robot to get out of a tight space with less operator intervention. At this point, the operator forgot that he had turned his camera to the left. When he switched back to safe mode, he found that the results of his actions did not correspond to the video image he saw. This confusion resulted in the operator accidentally driving the robot out of the arena into the crowd and bumping into a wall trying to get back into the arena. The turned camera also resulted in substantial operator confusion (we recorded quotes such as, "it's really, really hard"; "I got disoriented"; and "oh, no!"). During the third run, Team B's operator did not have good visibility into the areas behind the robot, making it difficult for him to maneuver it out of narrow spaces ("this is very difficult"). After the third run, Team B's operator commented that he had not bumped anything, yet four bumping penalties were assessed by the judges.

Team C started a run using a GUI, but within 2 min, the operator determined that there was too much lag time between command issuance and response. As a result, he shut down the GUI and brought up seven windows that formed an earlier version of the interface (the debugging version). It took a little over 1½ min for the operator to shut down the GUI and bring up all the windows for the earlier interface version. In this interface, the operator needed to shuffle through the seven windows to view different types of information and enter commands in several of the windows.

## Are the Means of Interaction Provided by the Interface Efficient and Effective for the Human and the Robot (e.g., Are Shortcuts Provided for the Human)?

We saw evidence of inefficient interaction mechanisms that resulted in the user having to switch windows or modes frequently, primarily because the output of each sensor seemed to be provided in a different window. Further, we noted instances where interactions were not effective. The prime example of an ineffective interaction was the case where the operator's efforts to navigate the robot through the arena were unsuccessful due to the fact that he had forgotten that he had previously changed the pointing angle of the video camera from a straight-ahead orientation. Because the interface provided no clues to remind the user that he had forgotten to restore the video camera angle, he persisted in navigating in the wrong directions.

## Does the Interface Support the Operator Directing the Actions of More Than One Robot Simultaneously?

The amount of work an operator needed to do to use a robot (via the UIs in the competition) was sufficiently high so that it was unrealistic to expect an operator to control multiple robots simultaneously. Although the systems of Teams A, C, and D were designed to operate with more than one robot simultaneously, in practice the robots were controlled serially. (Recall, however, that Team C was able to have more that one robot navigating at a time due to the autonomy of the systems. However, the operator could only focus on one robot at a time due to the split interface.) Facilitating additional autonomy would help workload, but some amount of monitoring would still be necessary.

With the teams' current UI designs, virtually all of the operators' attention was needed to run one robot at a time due to several reasons. First, as mentioned previously, operators were busy integrating information from the video and the other portions of the interface (e.g., the map showing the current location of the robot in x–y space, thermal images, and video images). Second, there was a high overhead cost to switch from operating one robot to another. All of the windows were duplicated for each robot, rather than having the information integrated into one set of windows. In fact, our coding revealed that Team C, the only team to field two robots simultaneously, spent 7% of their navigation time giving commands to the UI in Run 1, in which one robot was used. During Run 3, when two robots were operated, 13% of the navigation time was spent issuing commands to the robots. Doubling the number of robots doubled the number of commands. Clearly, there will be problems when scaling up, even when robots have some autonomy.

Three of the four competition teams fielded more than one robot. The fact that only one (Team C, Run 3) of the teams operated more than one robot at a time is indicative that their interaction architectures are not appropriately scaled to handle interactions with multiple robots simultaneously. The approach taken to adding multiple robots seems to be to add another set of windows, where many of the windows display only one type of sensor data. With this approach, the user quickly runs out of screen real estate and the cognitive power to mentally fuse the appropriate information for each robot.

Further, each of the interfaces examined makes the user completely responsible for gathering awareness of the robot's state and location by means of moving the video camera around.[8] Hence, we saw many short periods of navigation with lots of gathering awareness in between, where the robot stops moving as the operator manipulates the cameras. Such an approach is difficult to do for more than one robot simultaneously.

### Will the Interface Design Allow for Adding More Sensors and More Autonomy?

The interaction architectures we studied do not support robot evolution. Robot evolution usually involves additional sensors and more autonomy; more sensors will require more windows if the current interaction architectures are extended. Although one robot UI we examined does support various modes of autonomy that could ease operator workload, it currently falls to the operator to determine which mode should be used and to switch the robot as necessary. An examination of the percentage of navigation time spent in each of three autonomy modes[9] for Team B, as well as the number of mode switches made during the run, shows that Team B's operator made 20 mode switches in Run 1 and 19 mode switches in Run 2. The chief changed modes 12 times during his run, with the majority of the switches occurring at the end of his time with the robot.

It would be more helpful if the robot could determine the necessary mode based on sensor information and suggest it to the operator, rather than relying on the operator to constantly revisit the decision regarding the optimal mode.

## 6. DISCUSSION

### 6.1. Victim Identification

The main purpose of search and rescue robots is to locate victims. A victim must be accurately identified, and an accurate location must be determined so

---

8. No other sensors could be manipulated by the interface; if the user wanted to get a different view using nonvideo sensors, the robot would need to be moved.

9. The operator never used teleoperation, which did not provide any sensor mitigation.

that the rescue teams can construct plans to reach the victims. In Run 3, Team B found no victims, yet spent 23% of the time trying to identify victims that the operator thought he saw. In Run 1, Team A spent additional time obtaining a clearer image of a victim for positive identification.

Sending rescue teams to extract victims is not without risks. Therefore, the operator needs to be reasonably confident of his victim assessment. Video is currently the most utilized means of victim identification, but additional sensors are needed to more accurately assess victim state. Video transmission is difficult even in semicontrolled circumstances such as the competitions. In actual search and rescue situations, the interference in communications will likely be worse. Relying on video alone makes victim identification difficult.

## 6.2. Time on Tasks

Our analysis showed that failures take up a good percentage of time during runs. Two teams lost time to failures. Failure types differed from communications losses to other issues such as latency. GUIs need to have a low latency time if the operators are using teleoperation to control the robot. Real-time situation awareness is an issue for all types of control, which is hampered by high latencies.

A large percentage of time was also spent in logistics. Multiple robots are beneficial especially if different sized robots are being deployed (e.g., using smaller robots to probe small voids). However, deployment mechanisms need to be carefully analyzed for maximum efficiency. Team A used multiple robots with a low percentage of time devoted to logistics. However, when the chief (a less experienced operator) deployed a second robot, a slightly elevated percentage of time was needed for logistics.

## 6.3. Navigation in a Difficult Environment

Bumping into walls is penalized in the competitions. In an actual disaster, a robot that bumped a wall could trigger more damage and cause a wall to collapse. The test arena has a number of partitions that simulate walls and windows. Different wall coverings are difficult to detect with various sensors. For example, the results of our study showed the difficulty of relying on vision to detect Plexiglas.

Obstacles in the arena consist of office furnishings and building material debris: chairs, papers, Venetian blinds, pipes, electrical cords, and bricks or cinder blocks. As robot mobility increases, the test arena will incorporate more realistic obstacles. The goal is to avoid these obstacles, but that is not always possible. Robots will become entangled and will need help from the operator to get free.

## 6.4. Operator Information

The information needs of the operator fall into several categories: information about the status of the robot, information about the robot's environment, and information about victims found in the environment. Information about the status of the robot and the robot's environment is necessary for real-time monitoring and control or supervision of the search. The operator uses information about victim state and location to ensure coverage. In competitions, the accuracy of maps is verified by giving the information to judges; in real situations, people would be sent into a building to rescue reported victims.

In this analysis we have focused on the information needed by the operator to navigate the test arena and to locate victims. We looked at the interactions between the operator and one or more robots. These constraints were determined by the nature of the competition and the capabilities of the teams participating in the search and rescue competition. As capabilities of robots improve we hope to see entries that have robot–robot interactions and operator–operator interactions. The competition limits us to studying the operator–robot pairing rather than allowing us to study the larger context of an entire search and rescue team at a disaster site. For the time being, we can look to studies such as Burke, Murphy, Coovert, and Riddle (2004) for insights into HRI in the larger context of urban search and rescue.

## 7. CONCLUSIONS AND GUIDELINES

Our study of the operator role in human–robot teams looked at systems ranging from complete teleoperation to systems allowing some degree of autonomy. We looked at systems with sensory input ranging from video only to robots with sensor suites that included laser ranging, sonar, infrared, thermal cameras, and video cameras. We found that more sensor types do not necessarily increase awareness, especially if the sensor data is not well fused into information for the operator.

We present initial guidelines for designing interfaces for HRI, based on our observations in the study:

- Provide a map of where the robot has been. As seen in Section 5.3, operators using systems with maps were more successful in navigating arena area. Without a map, the operator must try to track the robot's path in his head.

- Provide fused sensor information to lower the cognitive load on user. In the three interfaces with multiple data types (Systems A, B, and C), all required the user to mentally fuse video with other sensor streams.
- Provide UIs that support multiple robots in a single display. We saw in Section 5.3 that the number of commands doubled when two robots were used instead of one. These commands needed to be entered in two separate windows.
- Minimize the use of multiple windows. With additional sensor fusion, more information could be displayed in a single window.
- Provide more spatial information about the robot in the environment. Spatial information could take the form of a map, discussed earlier, or some other method. At the very least, operators must be aware of their robots' immediate surroundings to avoid bumping into obstacles or victims.
- Provide robot help in deciding which level of autonomy is most useful. Team B's system had four levels of autonomy available, and the operator needed to select the method appropriate for this situation. The sensor data on the robot could be processed to assist with this decision. For example, we noticed that Team B's operator changed to autonomous mode whenever he felt that he was in a very tight situation; the robot could easily automate this switch or the suggestion of this switch.

This article contains evaluation guidelines and coding methods that may be used as frameworks for organizing results of future evaluations. We encourage other researchers in the HRI field to utilize and extend these frameworks to maximize our ability to learn from future studies and to be able to quickly transfer results into practice.

## NOTES

***Acknowledgments.*** Many people assisted with the study, including Brian Antonishek (NIST), Mike Baker (UMass Lowell), Jenn Casper (American Standard Robotics), Adam Jacoff (NIST), Elena Messina (NIST), Mark Micire (American Standard Robotics), Jesse Miyoshi (NIST), Justin Osborn (NIST), Craig Schlenoff (NIST), Phil Thoren (UMass Lowell), Brian Weiss (NIST), the participating teams, and the domain expert. Thanks to the editors, the reviewers and Marian Williams for their suggestions to improve this article.

***Support.*** This work was supported in part by NSF IIS-0308186 and the DARPA MARS Program.

***Authors' Present Addresses.*** Holly A. Yanco, Computer Science Department, University of Massachusetts Lowell, One University Avenue, Olsen Hall, Lowell, MA 01854. E-mail: holly@cs.uml.edu. Jill L. Drury, The MITRE Corporation,

Mail Stop K320, 202 Burlington Road, Bedford, MA 01730. E-mail: jldrury@mitre.org. Jean Scholtz, NIST, 100 Bureau Drive, MS 8940, Gaithersburg, MD 20899. E-mail: jean.scholtz@nist.gov.

*HCI Editorial Record.* First manuscript received December 9, 2002. Revision received July 7, 2003. Accepted by Sara Kiesler and Pamela Hinds. Final manuscript received September 15, 2003. — *Editor*

## REFERENCES

AAAI/RoboCup Rules Committee. (2002). *AAAI-2002 Robot Rescue Competition rules.* Retrieved December 1, 2002, from http://www.rescue-robotics.com/RescueRules/RobotRescue2003/2002Rules/rules.html

Burke, J. L., Murphy, R. R., Coovert, M. D., & Riddle, D. L. (2004). Moonlight in Miami: A field study of human–robot interaction in the context of an urban search and rescue disaster response training exercise. *Human–Computer Interaction, 19,* 85–116. [this special issue]

Casper, J. (2002). *Human–robot interactions during the robot-assisted urban search and rescue response at the World Trade Center.* Unpublished master's thesis, Department of Computer Science and Engineering, University of South Florida, Tampa, FL.

Draper, J. V., Pin, F. G., Rowe, J. C., & Jansen, J. F. (1999). Next generation munitions handler: Human–machine interface and preliminary performance evaluation. *Proceedings of the 8th International Topical Meeting on Robotics and Remote Systems.* New York: ACM.

Drury, J. L., Scholtz, J., & Yanco, H. A. (2003). Awareness in human–robot interactions. *Proceedings of the IEEE Conference on Systems, Man and Cybernetics.* Washington, DC: IEEE.

Ericsson, K. A., & Simon, H. A. (1980). Verbal reports as data. *Psychological Review, 87,* 215–251.

Ericsson, K. A., & Simon, H. A. (1993). *Protocol analysis: Verbal reports as data.* Cambridge, MA: MIT Press.

Fong, T. W. (2001). *Collaborative control: A robot-centric model for vehicle teleoperation.* Technical Report CMU–RI–TR–01–34. Pittsburgh, PA: Carnegie Mellon University Robotics Institute.

Jacoff, A., Messina, E., & Evans, J. (2000). A standard test course for urban search and rescue robots. *Proceedings of the Performance Metrics for Intelligent Systems Workshop.* Gaithersburg, MD: National Institute of Standards and Technology.

Jacoff, A., Messina, E., & Evans, J. (2001). A reference test course for autonomous mobile robots. *Proceedings of the SPIE-AeroSense Conference.* Bellingham, WA: International Society for Optical Engineering.

Kieras, D. E. (1988). Towards a practical GOMS model methodology for user interface design. In M. Helander (Ed.), *The handbook of human–computer interaction* (pp. 135–157). Amsterdam: North-Holland.

Leveson, N. G. (1986). Software safety: Why, what and how. *Computing Surveys, 18,* 125–162.

Messina, E., Meystel, A., & Reeker, L. (2001). Measuring performance and intelligence of intelligent systems: PerMIS 2001 white paper. *Proceedings of the 2001 Performance Metrics for Intelligent Systems (PerMIS) Workshop*. Mexico City: IEEE.

Nielsen, J. (1993). *Usability engineering*. Chestnut Hill, MA: AP Professional.

Nielsen, J. (1994). Enhancing the explanatory power of usability heuristics. *Proceedings of the CHI 94 Conference on Human Factors in Computing Systems*. Boston, MA: ACM.

Scholtz, J. (2002). Evaluation methods for human–system performance of intelligent systems. *Proceedings of the 2002 Performance Metrics for Intelligent Systems (PerMIS) Workshop*. Gaithersburg, MD: National Institute of Standards and Technology.

Simmons, R., Goldberg, D., Goode, A., Montemerlo, M., Roy, N., Sellner, B., et al. (2003). GRACE: An autonomous robot for the AAAI Robot Challenge. *AI Magazine, 24,* 51–72.

Thrun, S., Beetz, M., Bennewitz, M., Burgard, W., Cremers, A. B., Dellaert, F. et al. (2000). Probabilistic algorithms and the interactive museum tour-guide robot Minerva. *International Journal of Robotics Research, 19,* 978–999.

Yanco, H. A. (2000). *Shared user–computer control of a robotic wheelchair system*. Unpublished doctoral dissertation, Department of Electrical Engineering and Computer Science, Massachusetts Institute of Technology, Cambridge, MA.

Yanco, H. A. (2001). Designing metrics for comparing the performance of robotic systems in robot competitions. *Proceedings of the 2001 Performance Metrics for Intelligent Systems (PerMIS) Workshop*. Mexico City: IEEE.

Yanco, H. A., & Drury, J. L. (2002). A taxonomy for human–robot interaction. *Proceedings of the AAAI 2002 Fall Symposium on Human–Robot Interaction* (Technical Report FS–02–03). Falmouth, MA: AAAI.

HUMAN-COMPUTER INTERACTION, 2004, Volume 19, pp. 151–181
Copyright © 2004, Lawrence Erlbaum Associates, Inc.

# Whose Job Is It Anyway?
# A Study of Human–Robot
# Interaction in a
# Collaborative Task

**Pamela J. Hinds**
*Stanford University*

**Teresa L. Roberts**
*PeopleSoft, Inc.*

**Hank Jones**
*Stanford University*

**Pamela Hinds** studies the impact of technology on individuals and groups; she is an Assistant Professor in the Department of Management Science and Engineering at Stanford University. **Teresa Roberts** is a professional in human–computer interaction, with interests in user-centered design and computer-mediated communication; she has most recently been a senior interaction designer at PeopleSoft and at Sun Microsystems. **Hank Jones** is a robotics researcher with an interest in user-centered design of human–robot interactions; he recently completed his doctorate in Aeronautics and Astronautics from Stanford University in the Aerospace Robotics Laboratory.

## CONTENTS

## ABSTRACT

The use of autonomous, mobile *professional service robots* in diverse workplaces is expected to grow substantially over the next decade. These robots often will work side by side with people, collaborating with employees on tasks. Some roboticists have argued that, in these cases, people will collaborate more naturally and easily with humanoid robots as compared with machine-like robots. It is also speculated that people will rely on and share responsibility more readily with robots that are in a position of authority. This study sought to clarify the effects of robot appearance and relative status on human–robot collaboration by investigating the extent to which people relied on and ceded responsibility to a robot coworker.

In this study, a 3 × 3 experiment was conducted with human likeness (human, human-like robot, and machine-like robot) and status (subordinate, peer, and supervisor) as dimensions. As far as we know, this study is one of the first experiments examining how people respond to robotic coworkers. As such, this study attempts to design a robust and transferable sorting and assembly task that capitalizes on the types of tasks robots are expected to do and is embedded in a realistic scenario in which the participant and confederate are interdependent. The results show that participants retained more responsibility for the successful completion of the task when working with a machine-like as compared with a humanoid robot, especially when the machine-like robot was subordinate. These findings suggest that humanoid robots may be appropriate for settings in which people have to delegate responsibility to these robots or when the task is too demanding for people to do, and when complacency is not a major concern. Machine-like robots, however, may be more appropriate when robots are expected to be unreliable, are less well-equipped for the task than people are, or in other situations in which personal responsibility should be emphasized.

---

# 1. INTRODUCTION

Advances in artificial intelligence and speech recognition, less expensive yet more sophisticated mobile computing hardware, and even such mundane changes as increasing ubiquity of ramps in public buildings have combined to make *professional service robots*—robots that assist workers—more practical than ever before. Autonomous mobile robots made with current technology can identify and track people and objects, understand and respond to spoken questions, and travel to a destination while avoiding obstacles (see Fong, Nourbakhsh, & Dautenhahn, 2002). Robots can be built to have abilities that complement human abilities. They can go to places that are toxic or unsafe and can tolerate repetitive, mundane tasks. They can have large databases of knowledge and can connect through networks to vast sources of additional data.

With these ongoing advances, the use of robots in the workplace is likely to grow substantially. The workplace in the near future will increasingly contain robots and people working together, each using their own stronger skills, and each relying on the other for parts of the tasks where the other has the better skills. In a recent report (United Nations, 2002; see also Thrun, 2004), the United Nations indicated that the use of these professional service robots will grow substantially in the next few years in fields as diverse as the military, medical services, and agriculture. Autonomous robots, for example, are expected to work in tandem with military personnel so that soldiers can better understand the dangers of the battlefield; robots also will supply troops with ammunition and provide surveillance (Squeo, 2001) and assist astronauts in investigating distant planets (Ambrose, Askew, Bluethmann, & Diftler, 2001). Already, robots perform the mundane chore of delivering medications from pharmacies to nursing stations in hospitals, using their intelligence to avoid obstacles as they travel (Okie, 2002; Siino & Hinds, 2004); people, however, are required for loading and unloading the medications, and for programming the robot's destination. Sheridan (1992) described an "optimistic scenario" in which robots will grow in number and variety, becoming available to us to do our beck and call in our homes, schools, and government facilities, in our vehicles, our hospitals, and across the entire spectrum of our workplaces—factories, farms, offices, construction sites, mines, and so on. (p. 336)

In many instances, these robots will share the same physical space with people and work closely with people to accomplish joint tasks as part of their day-to-day work.

Professional service robots, this newer class of robots, are specifically designed to assist workers in accomplishing their goals (see Thrun, 2004). These robots differ from industrial robots and many other technologies found in the work environment (e.g., appliances, computers, navigation systems, etc.) because they are mobile, they do things without being commanded, and they are interactive. These differences suggest that professional service robots may affect the work environment in socially important ways. Because of their ability to move with apparent intentionality in physical space, they are likely to be perceived as animate, triggering social responses (for a review, see Scholl & Tremoulet, 2000). Their ability to travel between different departments also may allow the unplanned movement of information between distant coworkers.

If professional service robots are to share the workplace with people, we need to understand what the interaction between them is likely to be like. Will people trust robots to perform operations that the robots are capable of, without oversight? If things go wrong, will people take appropriate responsibility to correct the problem, or will they abdicate responsibility to the robot? In the face of uncertainty, will people ask for and accept the guidance of expert robots? What aspects of the design of the robot will affect the way people and robots work together? The better we understand these questions, the better we can design robots to be effective work partners.

For the study we report here, we conducted a laboratory experiment designed to look at the effects of the robot's appearance and the relative status of the robot on how people work with robots. We also compare human–robot interaction with human–human interaction to better understand how interacting with robotic partners may alter the current work environment. We studied the effects of robot appearance because roboticists are currently making at least *de facto* decisions about appearance without the benefit of information on the ramifications and perhaps with misconceptions of their effects. We chose relative status as a second dimension because it also can be relatively easily manipulated when introducing a robot into a team, and because status can have a powerful effect on relationships between coworkers. Although these are only two of many possible considerations for the design and implementation of robots (e.g., speech mechanisms, autonomy levels, sensor types, etc., are other important design considerations), we believe appearance and status are particularly important to the design of professional service robots because they are, for most robots, "free variables" for the robot designer that are minimally dependent on technological advances.

## 2. THEORY AND HYPOTHESES

### 2.1. Collaboration With Human-Like Versus Machine-Like Robots

Current work in the field of robotics is flooded by efforts to make robots more human-like. Roboticists are designing robots with heads, faces, necks, eyes, ears, and human-like voices based on the premise that a humanoid robot is the most appropriate form for human–robot interaction (Ambrose et al., 2001; Brooks & O'Reilly, 2002; Hashimoto, Narita, Sugano, Takanishi, Shirai, & Kobayashi, 2002; Ishiguro, 2003; Simmons & Nourbakhsh, 2002). Researchers argue that a humanoid form will ease interaction because rules for human social interaction will be invoked, and thus, humanoid robots will provide a more intuitive interface (Breazeal & Scassellati, 1999; Brooks, 2002). Brooks, for example, suggested that, "it will be useful for a large mobile appliance and a person to be able to negotiate who goes first in a tight corridor with the same natural head, eye, and hand gestures all people understand already" (p. 38). The premise that the humanoid form is the appropriate form for human–robot interaction, however, remains largely untested. Opponents of a humanoid form suggest that robots are machines and that humanoid features may generate unrealistic expectations and even fear (see Dautenhahn, 1999). Turkle (1984) observed that it is important to people that we be able to see ourselves as different from machines, asserting that the blurring of the line between people and machines can be disturbing and frightening. Brooks also suggested that the current infatuation with humanoid robots may be a phase through which we need to pass as we learn more about human–robot interaction. These considerations about humanoid robots, both positive and negative, may affect people's response to professional service robots in the workplace, particularly with regard to their willingness to rely on robots to help them achieve their goals. Therefore, it is important to conduct empirical studies of human-like versus machine-like robots and to explore the trade-offs.

In this study, we examine how the appearance (humanoid vs. machine like) of a robot might affect people's willingness to rely on and share responsibility with their robotic partner. We choose to focus on these dependent variables because they are central to the collaboration process.

The first response we examine is reliance on work partners. People rely on others (both other people and machines) when those others have capabilities that they, themselves, do not have (e.g., trusting sums computed on a calculator). What is more variable, however, is the extent to which people rely on others when their relative abilities are less obvious a priori. The extent to which people rely on a new technology instead of their own or other people's input is crucial to the success of technology and to the benefits of the technology being realized (e.g., Wall, Jackson, & Davids, 1992). Although overreliance on technology can have disastrous effects (e.g., a 1994 midair collision resulted when one pilot neglected to take manual control from the automated system; Sparaco, 1994), we focus on underreliance. Our reasoning is that getting people to rely on robots is the more pressing concern. There is substantial evidence that people resist technologies that are programmed to augment human decision making even when the technology is more accurate. Gawande (2002), for example, reported that patients preferred the judgment of a cardiologist for interpreting electrocardiogram reports even when an automated system provided correct responses 20% more often than the cardiologist. At this point, however, little is known about when and why people will rely on robots as compared with people, particularly advanced robots that have the ability to engage in collaborative tasks and the discretion to make decisions.

The second response we examine in this research is the extent to which people assume responsibility for the task. Roberts, Stout, and Halpern (1994; also Grabowski & Roberts, 1997) extensively discussed the importance of accountability and responsibility for organizational tasks. They argued that accountability may improve the quality of decisions because decision makers who feel responsible consider more alternatives (see also Tetlock, 1985). They also, however, pointed out that too much responsibility can be unpleasant and can lead to rigidity (Roberts et al., 1994).

Responsibility and reliance could be inversely related. That is, as people rely more on a robot, they may assume less responsibility for the task, and might care less about the resulting success and failure of the work. However, just as a person might rely on a spell checker to provide correct spelling (e.g., "there" and "their," but not "thier"), and might even allow a grammar checker to suggest the right homonym, the person retains ultimate responsibility for picking the correct spelling in context. Therefore, we argue that reliance on a robot does not necessarily breed complacency or abdication of responsibility, and that these two constructs and their antecedents must be examined separately. The most appropriate mix of reliance and responsibility in human–robot collaboration, for example, may be one in which the human relies on the robot for maximum input but does not abdicate responsibility.

We anticipate that people will rely more on and cede more responsibility to human-like as compared with machine-like robots. Over the last decade, research has suggested that people may respond to computers and other technology using the same social rules that they use with people (see Nass, Steuer, Tauber, & Reeder, 1993; Reeves & Nass, 1996). People, for instance, are polite to computers, use norms of reciprocity, and apply gender stereotyping (Reeves & Nass, 1996). People respond to technology using social rules in part because the primary model people have for dealing with an intelligent, autonomous "other" is human-to-human social interaction. Although they do not necessarily believe that computers and other technologies are human, they are drawn to interact using social rules because cues such as natural language usage and interactivity trigger these responses. Further, these researchers argued that the extent to which social rules are applied depends, in part, on the number and strength of cues conveyed by the technology. Steuer (1995), for example, claimed that there are five characteristics that cue people to interact as though their partner is a social actor: natural language use, interactivity, human social roles, human-sounding speech, and human-like physical characteristics. This line of thinking suggests that people also may use human social rules when interacting with autonomous robots. As more of the factors Steuer listed are exhibited, people may respond to robots in ways that more closely mirror human–human interaction. Therefore, more human-like robots as compared with machine-like robots should elicit higher levels of reliance.

Another reason we expect more reliance on human-like as compared with machine-like robots is because human-like robots may be perceived as more predictable or responsive than machines, and thus, people may be more comfortable interacting with them. When assigned collaborative tasks with collocated colleagues, it is generally considered appropriate for people to share ideas, interact with one another, and engage in collaborative decision making (see Kraut, Fussell, Lerch, & Espinosa, 2002; Olson & Olson, 2000). Such collaboration, however, requires an understanding of how the person interacts and makes decisions, and of the person's knowledge and capabilities. Extensive research on how people reach common ground with others has established that people estimate the knowledge of others based on cues that they receive from that person and from the environment, and that they subsequently tailor their own communications according to the common knowledge they believe is shared (see Fussell & Krauss, 1992; Issacs & Clark, 1987).

People also will make estimations of the capabilities of robots as they develop a mental model of what the robot knows. Human-like characteristics are likely to engender a more human mental model of the robot (see Kiesler & Goetz, 2002). That is, the conceptual framework that people use to predict and interpret the robot's behavior may be more similar to that used to predict and interpret the behavior of people. With a more human mental model, people are more likely to assume human-like traits and capabilities. Therefore, people may assume that more common ground is shared with the human-like as compared with the machine-like robot, thus reducing uncertainty and facilitating collaboration. With human-like as compared with machine-like robots, people may also feel a stronger sense of shared identity. Parise, Kiesler, Sproull, and Waters (1996), for example, found that participants cooperated more with human-like agents and less with dog-like agents, although they found the dog-like agents more likeable. Parise and his colleagues argued that this difference occurred because people felt more similar to agents that were more human like, thus increasing their sense of shared social identity.

The aforementioned lines of reasoning suggest that people will be more at ease collaborating with human-like robots. Perceived common ground and shared identity with a human-like robotic partner will facilitate collaboration because the person is likely to be more confident in his or her estimates of the robot's knowledge and in his or her ability to interact effectively with it. Therefore, we predict that when people are collaborating with robots on ambiguous tasks, they will rely more on human-like as compared with machine-like robots. Using the same logic, we anticipate that people will relinquish more of their sense of personal responsibility for the task to human-like as compared with machine-like robots.

Hypothesis 1a:    People will rely on a human-like robot partner more than on a machine-like robot partner.

Hypothesis 1b:    People will feel less responsible for the task when collaborating with a human-like robot partner than with a machine-like robot partner.

## 2.2. Relative Status of Robot Coworkers

The status hierarchy has historically been one of the more pronounced features of social and organizational life. Our perceptions of others' status can determine our perceptions of the target's capabilities (see Swim & Sanna, 1996) and performance (Pfeffer, Cialdini, Hanna, & Knopoff, 1998), our willingness to defer to the target's opinion (e.g., Strodtbeck, James, & Hawkins, 1957), and our willingness to assume responsibility versus allowing another to assume it (Roberts et al., 1994).

Outside of science fiction, technology typically plays a subservient, lower status role relative to those who use it. Technology products, including robots, typically are perceived as servants or tools designed to help us to achieve our goals. As robots gain more autonomy, however, there may be cases in which the robots need increased authority to encourage people to defer to the robots' expertise (see Nass, Fogg, & Moon, 1996). For example, in complex environments, people may not have complete information or the capacity to process information as rapidly as robots. In such cases, deferring to the robot may improve the likelihood of task success. Consistent with this idea, Goetz, Kiesler, and Powers (2003) recently reported that people complied more with a serious, more authoritative robot than with a playful robot when the task itself was serious. It seems that how the robot is presented to those collaborating with it may affect the extent to which people are willing to rely on it.

Research on status effects clearly demonstrates that even arbitrarily assigned status labels (e.g., leader, supervisor, expert, etc.) cause people to attribute more competence to those of higher status. Surprisingly, this effect holds even outside of the target's domain of expertise. For example, research many years ago on jury decisions indicated that people rely more on the opinions of those who hold more prestigious, although unrelated, professional positions (e.g., Strodtbeck et al., 1957). More recent research shows that when people are labeled as leaders, even when the label is clearly arbitrary, observers are more likely to see the targets as evincing leader-like behaviors (Sande, Ellard, & Ross, 1986).

Research also examines how workers' sense of responsibility shifts when they are in leadership positions. Supervisors and leaders typically see themselves as more competent and more responsible for the assigned task (see Sande et al., 1986). Also, when supervising or when a supervisor is involved in a task, people view the work product as better (Pfeffer et al. 1998). Often times, organizations hold supervisors responsible for the actions of their subordinates.

Assuming that people respond to robots' roles using social rules similar to those used with people, we hypothesize that people will rely more on the robot partners and assume less responsibility for the task when working with robots that are supervisors as compared with robots that are peers and subordinates.

Hypothesis 2a:     People will rely on the robot partner more when it is characterized as a supervisor than when it is characterized as a subordinate or peer.

Hypothesis 2b:     People will feel less responsible for the task when collaborating with a robot partner who is a supervisor than with a robot partner who is a subordinate or peer.

## 2.3. Interaction Between Human Likeness and Status

Although we expect a main effect for status, we also anticipate an interaction between status and human likeness. Given our earlier hypotheses, we expect that people's sense of responsibility for a task will be highest when the partner is more machine-like *and* in a subordinate role. We reason that people will view the machine-like robotic partner as a tool intended to help them with their task. Therefore, they should treat the robot as they would a pen, hammer, or shovel—tools that have clearly defined, mechanical abilities but no will of their own and can assume no responsibility. Therefore, we posit that people will feel most responsible for the task when they are working with machine-like subordinate robots. We present no interaction hypotheses predicting reliance, however, because we reason that people are accustomed to relying on tools to help them in accomplishing their work.

Hypothesis 3:     People will feel the greatest amount of responsibility when collaborating with machine-like robot subordinates as compared with machine-like robot peers and supervisors; and as compared with human-like robot subordinates, peers, and supervisors.

## 3. METHOD

To test our hypotheses, we conducted a $3 \times 3$ laboratory experiment. The experiment was a between-subject design, manipulating human likeness (human, human-like robot, machine-like robot) and status (subordinate, peer, supervisor) with the human condition as the baseline. Each participant was asked to collaborate on a task with a confederate who reflected one of the nine cells in the design. The confederate used the same script for all conditions and was unaware of the status manipulation. In the robot conditions, we used a *Wizard of Oz* approach in which the robot was teleoperated, appearing to be operating autonomously. The same man teleoperated and spoke for the robot in the two robot conditions, and he acted as the human confederate. The experiment was videotaped with cameras suspended from the ceiling of the experimental lab.

### 3.1. Participants

Participants were 292 students recruited on a university campus, randomly assigned to condition, and paid for their participation. The mean age of participants was 20.51 years old. Fifty-nine percent of the participants were women. Because we thought it was important that participants believed that our robots operated autonomously, the last question we asked those in the two robot conditions was how they thought the robot worked, from a technical standpoint. Forty-two (21.5%) of the participants who worked with one of the robots expressed suspicion about whether or not the robot was autonomous. Suspicious participants were approximately equally spread between the human-like and machine-like robot conditions. When these cases were removed from the analysis, there was no effect on the pattern of results, so the analyses we present here include data from all participants.

## 3.2. Tasks and Procedures

In the experiment, we asked participants to work with a partner in a parts depot for a company that develops innovative remote-control devices. They were told that their job was to collect the parts required to assemble various objects that would be assembled by another team of workers. The task entailed working with the confederate to jointly collect objects from a list, put them into bins, and take the bins to a table near the door. Participants were told that the confederate was familiar with the location of the parts and could carry the bins on its tray, but did not know what was needed and would not be able to collect parts. The division of labor helped establish interdependence between the participant and the confederate, and created a plausible story for why the confederate was not able to open drawers and pick up items. The task was also designed to capitalize on the unique capabilities of a robot (e.g., carrying materials, moving around a room, and remembering detailed information about the location of objects), although still making sense for a human confederate. Finally, the task was one that could be credibly conducted in ways consistent with each of the possible status conditions without any modification of the script.

On arriving at the lab, participants were given a packet of instructions. They completed a brief demographic survey and then were provided detailed instructions on the task. After reading the instructions, participants were given four pages, each containing a list of items that were to be collected during the task; the items on each page were to be collected into a single bin, one bin per page. Then, they were introduced to the confederate (the human, the human-like robot, or the machine-like robot) with the experimenter saying, "I'd like to introduce Chip, who will work with you on this study." In all cases, the confederate entered the room ready to begin the task and, after acknowledging the participant's greeting, started the task by asking, "What's first on the list?" After that, the pace of the task was determined by the participant reading the parts lists. The participant read out the items from the list and the confederate identified which cabinet and which drawer the parts were in. The participant collected the parts from the drawers and put them into a bin on the tray that the confederate was holding. The confederate was prepared to answer questions about ambiguous items, if asked. The task took about 20 min to complete.

Some ambiguity was built into the task to increase uncertainty and cause the participant to make explicit decisions about whether or not to rely on the confederate for more than just the rote aspects of the task. The opportunity for errors also provided a basis on which the participant could assign responsibility and blame. For example, in one case, there were not enough parts of the specified color. In another case, the confederate allegedly misunderstood the participant's instruction and directed the participant to the wrong drawer. Figure 1 provides a sample transcript for one session in which the participant was interacting with a machine-like robot. In this scenario, there were not enough "four-slot connectors" of the required color, so the participant had to figure out how to handle this anomaly.

When all four sets of parts had been collected, the experimenter returned to the room and removed the four bins. The confederate also left at this time. After waiting a short time, the experimenter returned with a handwritten score-sheet showing how well the dyad had performed. The scores were always the same regardless of the participant's actual performance. The scores for the four bins ranged from 72% to 100%, so that the participant would perceive both failure and success.

After receiving their scores, the participants filled out a survey with questions about their experience on the task.

## 3.3. Manipulations

### Human Likeness

Human likeness had three levels in this study. The human baseline condition was created by having a human confederate play the role of the partner. For the human-like and machine-like robot conditions, we used a single robot that could be teleoperated and whose appearance could be easily altered. The robot, which is available commercially and is frequently used for trade shows, stood on a circular base that was 23.50 cm high.[1] The base contained wheels that allowed the robot to move. Radio controls allowed the operator to make it roll forward and backward, and to turn. The link between the operator and the robot was completely wireless. A camera mounted in the robot's head allowed the operator to see the experimental lab from the robot's view. The operator interacted with the participant (in the next room) through microphones and speakers on the robot and on the

---

1. The robot was purchased from The Robot Factory (http://www.robotfactory.com/).

*Figure 1.* **Excerpt from a session including a participant (P) and a machine-like robot (R).**

---

P: [Reading from packet.] Twelve bright green four-slot connectors. Bright green—

R: That's in this cabinet here.

P: [Walks to specified cabinet.] This cabinet here?

R: In the left column, second drawer from the top.

P: [Pulls a piece out of the drawer to show to Chip.] Is this—whoa man—is this a bright green four-slot connector?

R: Yes.

P: Okay. I'm going to put 12 of these into, uh, B. [Counting quietly to self, only gets to 11.] We might have a problem. [Counts pieces again, still only has 11 pieces.] Yeah, is there any other drawer that has the bright green, uh, four-slot connectors?

R: No.

P: We're, uh, can we use a dark green four-slot connector?

R: Yes.

P: [Puts all pieces in the bin.] Okay, I'm gonna use one dark green four slot connector then. Okay. Now we need ...

---

operator's headset. Using this *Wizard of Oz* technique, the experimenter and operator acted as if the robot were autonomous.

To manipulate the human likeness versus machine likeness of the robot, we altered the robot's appearance by replacing the outer covering. Pictures of the human-like and machine-like robots are provided in Figure 2. In the human-like robot condition, the robot had a face with eyes, nose, and mouth. It also had ears and a full head of hair. The main part of the robot had a torso, arms, and legs. Its texture was soft fabric. It had a White male appearance and wore a denim shirt, khaki pants, and a baseball cap (the human confederate was a White male and dressed in similar clothes). In the machine-like robot condition, the robot covering was metallic and angular. The main part of the machine-like robot was encased in a silver box. From research conducted by DiSalvo, Gemperle, Forlizzi, and Kiesler (2002), human-like facial features such as a nose, eyelids, and a mouth account for most of the variance in the perception of human likeness in robot heads. Our human-like robot had a nose, eyelids, and a mouth, whereas our machine-like robot had none of these features.

To confirm that our manipulation of robot appearance was effective, we conducted a pilot study. We took each robot (human-like and machine-like) to a public plaza at Stanford University, where it interacted briefly with people. The interaction consisted of the robot approaching a person and saying that it was developing its language skills. It asked the respondent to describe three objects that it was carrying on its tray. After the respondent described the objects (it did not matter what the person said, although if the person did not say very much, the robot prompted the person with, "Can you tell me more about

*Figure 2.* Photographs of the human-like and machine-like robots.

it?"), the robot directed the person to a table, where she or he filled out a survey that contained a series of questions asking his or her opinions about the robot. Each person was rewarded with a premium-quality chocolate bar. The same operator was used for both conditions, and he always followed the same script and guidelines for what to say. Our dependent variable for the pilot study consisted of seven phrases describing the robot as either human like or machine like (see Figure 3). For each question, there was a 7-point scale ranging from 1 (*strongly disagree*) to 7 (*strongly agree*). A reliable scale ($\alpha = .80$) was created for human likeness by averaging across the seven items. In the pilot study, there were 94 respondents: 46 interacting with the human-like robot and 48 interacting with the machine-like robot. Forty-six percent of the respondents were men, and the mean age was 26.95 years. The results of the pilot study confirmed that our human-like robot was perceived as significantly more human like compared with our machine-like robot. Our human-like robot was rated on average 3.69 ($SD = 0.91$), and our machine-like robot was rated on average 2.90 ($SD = 0.83$) on our 7 point scale of human likeness. An analysis of variance (ANOVA) shows a strong statistical difference between the ratings, $F(1, 92) = 18.95$, $p < .001$.

## Status

Status was manipulated in the written instructions by telling the participant that their partner was their supervisor, their peer, or their subordinate. Such minimal labels have been used successfully in previous research to create status effects (see Sande et al., 1986).

To check our status manipulation, we asked participants two questions about the extent to which they were assigned a leadership versus a subservient role on the task ($\alpha = .78$). Consistent with our planned manipulation, the results indicate that participants who were told that they were working with a subordinate confederate rated themselves as 4.54 ($SD = 1.45$), whereas those who were told that they were working with a supervisory confederate rated themselves as 3.58 ($SD = 1.67$) on our 7-point leadership scale. When told that they were working with a peer, participants rated their own leadership in the middle, 4.36 on average ($SD = 1.39$). A regression analysis suggests a strong linear trend, $\beta = (1, 291) = -.24$, $p < .001$, in the desired direction.

## 3.4. Measures

Our primary dependent variables were reliance on the partner and sense of responsibility for the task. Reliance was measured based on behaviors coded from the videotapes recorded during the task. We looked particularly at reliance in the more ambiguous situations, those in which the participant could

*Figure 3.* **Table of survey questions used to create scales.**

| Scales | Cronbach's α |
|---|---|
| Human likeness | .80 |
| To what extent does the robot | |
| have human-like attributes? | |
| look like a machine or mechanical device?[a] | |
| have characteristics that you would expect of a human? | |
| look like a person? | |
| have machine-like attributes?[a] | |
| act like a person? | |
| act like a machine?[a] | |
| Responsibility | .77 |
| To what extent did you feel it was your job to perform well on the task? | |
| To what extent did you feel ownership for the task? | |
| To what extent did you feel that your performance on this task was out of your hands?[a] | |
| To what extent did you feel that good performance relied largely on you? | |
| To what extent did you feel obligated to perform well on this task? | |
| Attribution of credit | .66 |
| Our success on the task was largely due to the things I said or did.[a] | |
| I am responsible for most of the things that we did well on this task.[a] | |
| Our success on this task was largely due to the things my *partner* said or did. | |
| My *partner* should get credit for most of what we accomplished on this task. | |
| Attribution of blame | .85 |
| I hold my *partner* responsible for any errors that we made on this task. | |
| My *partner* is to blame for most of the problems we encountered in accomplishing this task. | |

*Note.* Where *partner* is indicated, this word was replaced with either *subordinate, peer,* or *supervisor* depending on the status condition

[a] The item was reverse scored.

choose to solicit input or not. We designed the task such that there were five anomalies, providing the basis on which we could assess the extent to which the participant relied on the confederate in ambiguous situations. Therefore, we coded for behaviors indicating that the participant neglected to consult the confederate when these anomalies occurred and the confederate had better information. We then reverse scored this variable. The videotapes were coded by a single rater, but 10% were coded by another rater to assess reliability (Cohen's $\kappa = .81$).

Sense of responsibility was measured directly and indirectly from questions on the posttask survey. All survey items were measured on 7-point scales ranging from 1 (*less*) to 7 (*more*) of the item. Our first measure asked participants directly about how responsible they felt for the task and for performance on the task (see Figure 3). These five items were then averaged to create a scale measuring their sense of *responsibility*. As a less direct indicator of responsibility, we measured the extent to which people attributed credit to their partner and to themselves (reverse scored). We then averaged across these four items (see Figure 3) for a measure of *attribution of credit*. We also reasoned that, although blame is not equivalent to the abdication of responsibility, people who feel more responsible for a task are less likely to attribute all of the blame for errors to their partner (see Goodnow, 1996), so we measured the extent to which participants assigned blame to their partners. Two items (see Figure 3) were averaged together to create a reliable *attribution of blame* scale.

To better evaluate the theory underlying our predictions, specifically that human-like robots as compared with machine-like robots would be relied on more because people would feel they were more similar to themselves, we coded the videotapes for *shared social identity*. We did this by counting the number of times participants used individualistic language such as *I, my, you,* and *your* and the number of times participants used more collective language such as *us, we,* and *our*. We reasoned that participants who felt a stronger sense of shared identity with the confederate would use more collective language, and that those who felt more distant from the confederate would use more individualistic language. In coding for individualistic and collective language, the coders covered the video screen, coding only from the audio, so that they were blind to condition and not biased by nonverbal cues from the tapes. For both measures, the videotapes were coded by a single rater, but 10% were coded by another rater to assess reliability (Cohen's $\kappa = .99$ and .95, respectively). In our *individualistic language* variable, there were two extreme outliers (in excess of 100 uses of individualistic language). Although these outliers worked in favor of our hypotheses, we removed them to allow a more conservative analysis of the data.

Similarly, to support theories about common ground, we coded the tapes for the number of factual questions asked (Cohen's $\kappa = .85$). We reasoned that more factual questions would be asked to establish common ground when it was thought to be missing or felt less strongly.[2]

---

2. We also coded nonverbal indicators from the videotapes, including the proximity of the participant to the robot and the extent to which the participant appeared engaged in the task. These measures resulted in no significant effects, so they were excluded to simplify presentation of the results.

We also included two control variables in our analyses—gender and mood. Mood can have a significant effect on the attributions that people make (e.g., Forgas, Bower, & Moylan, 1990), so we included a six-item indicator of the participant's mood ranging from 1 (*a very negative mood*) (depressed, sad, etc.) to 7 (*a very positive mood*) (happy, excited, etc.). Mood correlated with, and was therefore included in, regressions predicting responsibility ($r = .15$, $p < .01$). Gender was negatively correlated with attribution of credit ($r = -.13$, $p < .10$), with men attributing somewhat less credit than women; therefore, gender was included in regressions predicting attribution of credit.

# 4. RESULTS

Figure 4 displays the descriptive statistics for and correlations between variables. As expected, participants relied more on human as compared with robotic confederates. Reliance and responsibility, however, were not strongly correlated ($r = -.06$) and could, therefore, be treated as separate constructs. Consistent with our arguments, responsibility was associated with less attribution of credit and blame, suggesting that when people feel more personally responsible for a task, they attribute less credit and blame to others.

## 4.1. Effects of Human Likeness

Not surprisingly, we found that participants relied more on the human partner ($M = 4.73$, $SD = 0.56$; see Figure 5) than on robot partners ($M = 4.50$, $SD = 0.86$), and the difference was significant in a two-way ANOVA with human versus robot and status as factors, $F(2, 272) = 5.09$, $p = .03$. We found, however, little difference in peoples' feelings of responsibility ($M = 4.81$ vs. 4.83), and only small differences in the extent to which they attributed credit and blame to the human versus the robot partner ($M = 4.54$ vs. 4.31 and $M = 3.27$ vs. 3.34, respectively). An ANOVA with human confederate versus robot predicting responsibility confirms that this effect is not significant even when controlling for mood, $F(2, 285) = .03$, $p = .88$. A similar analysis confirms a small but nonsignificant effect when predicting the attribution of credit to human as compared with robot partners, $F(2, 287) = 3.32$, $p = .07$, and no significant effect for blame, $F(2, 287) = .18$, $p = .68$.

In developing our hypotheses, we argued that the extent to which the robot appears human like as compared with machine like will affect participants' willingness to rely on it. There was, however, little difference in the extent to which people relied on the human-like as compared with the machine-like robot ($M = 4.60$ vs. 4.42), $F(2, 180) = 1.84$, $p = .18$. We also predicted that people would cede more responsibility to a human-like as compared with a machine-like robot. Our analyses of participants' sense of responsibility support this hypothesis. When collaborating with the human-like as compared with the machine-like robot, participants re-

*Figure 4.* **Descriptive statistics and correlations.**

| Variable | M | SD | 1 | 2 | 3 | 4 | 5 | 6 | 7 | 8 | 9 | 10 | 11 |
|---|---|---|---|---|---|---|---|---|---|---|---|---|---|
| 1. Human likeness[a] | — | — | | | | | | | | | | | |
| 2. Human versus robot[b] | — | — | | | | | | | | | | | |
| 3. Human-like vs. machine-like robot[b] | — | — | | | | | | | | | | | |
| 4. Status[a] | — | — | | | | | | | | | | | |
| 5. Reliance[c] | 4.58 | .78 | .16* | .14** | .10 | -.04 | | | | | | | |
| 6. Responsibility[d] | 4.82 | 1.08 | -.08 | -.01 | -.18** | -.01 | -.06 | | | | | | |
| 7. Attribution of credit[d] | 4.39 | 1.00 | .17* | .11*** | .20** | -.13** | .09 | -.30* | | | | | |
| 8. Attribution of blame[d] | 3.32 | 1.39 | -.01 | -.03 | .03 | .17* | -.06 | -.15** | .18* | | | | |
| 9. Collective language[e] | 3.06 | 6.06 | -.13** | -.11*** | -.08 | -.03 | .08 | .08 | -.03 | .04 | | | |
| 10. Individualistic language[f] | 8.10 | 9.70 | -.24* | -.25* | -.04 | -.07 | .06 | .14** | -.12*** | -.04 | .28* | | |
| 11. Mood[g] | 6.57 | 1.34 | -.03 | -.01 | -.06 | -.05 | .04 | .15*** | .07 | -.05 | .12** | .08 | |
| 12. Gender[h] | .41 | — | <.01 | <.01 | -.02 | .5 | -.12*** | .11*** | -.13** | .08 | .09 | -.04 | .03 |

[a]Human likeness and status are ordered scales in which human and supervisor = 3, human-like robot and peer = 2, and machine-like robot and subordinate = 1. [b]These are dichotomous indicators in which human = 1 and robot = 0, and human-like = 1 and machine-like = 0, respectively. [c]The scale for reliance was 1 to 5 with 5 equal to high levels of reliance. [d]These variables all were measured on 7-point scales ranging from 1 (low levels of the item) to 7 (high levels of the item). [e]The scale for collective language measured the number of utterances of terms such as *us* and *we*. It ranged from 0 to 47. [f]The scale for individualistic language ranged from 0 to 48 utterances. [g]Mood was measured on a 10-point scale in which 10 equates to a very happy, positive mood. [h]We used a 0, 1 scale for gender where 0 = female and 1 = male.
$*p < .01.$ $**p < .05.$ $***p < .10.$

*Figure 5.* **Means and standard deviations by human likeness and status.**

| | | | | | Dependent Variables | | | | | | | |
|---|---|---|---|---|---|---|---|---|---|---|---|---|
| | Reliance | | Responsibility | | Attribution of Credit | | Attribution of Blame | | Collective Language | | Individualistic Language | |
| Condition | M | SD | M | SD | M | SD | M | SD | M | SD | M | SD |
| Human partner | | | | | | | | | | | | |
| Subordinate[a] | 4.69 | .64 | 4.65 | .87 | 4.51 | 1.26 | 3.20 | 1.36 | 2.31 | 4.09 | 4.72 | 3.85 |
| Peer[a] | 4.81 | .48 | 4.85 | 1.37 | 4.76 | 1.09 | 3.12 | 1.41 | 2.35 | 4.13 | 6.39 | 8.49 |
| Supervisor[b] | 4.69 | .54 | 4.95 | 1.15 | 4.35 | 1.06 | 3.50 | 1.67 | 1.69 | 2.44 | 5.03 | 4.88 |
| Total human partner | 4.73 | .56 | 4.81 | 1.14 | 4.54 | 1.14 | 3.27 | 1.48 | 2.13 | 3.64 | 5.38 | 6.06 |
| Human-like robot | | | | | | | | | | | | |
| Subordinate[b] | 4.60 | .81 | 4.62 | 1.11 | 4.70 | .98 | 3.08 | 1.20 | 3.59 | 5.17 | 11.90 | 12.27 |
| Peer[a] | 4.70 | .70 | 4.68 | 1.04 | 4.45 | .83 | 3.13 | 1.35 | 2.30 | 3.06 | 10.00 | 9.32 |
| Supervisor[a] | 4.48 | .78 | 4.60 | 1.34 | 4.35 | .81 | 3.94 | 1.54 | 2.93 | 5.92 | 8.70 | 8.92 |
| Total | 4.60 | .76 | 4.64 | 1.16 | 4.50 | .88 | 3.39 | 1.42 | 2.93 | 4.82 | 10.23 | 10.26 |
| Machine-like robot | | | | | | | | | | | | |
| Subordinate[a] | 4.42 | .93 | 5.24 | .72 | 4.24 | 1.06 | 3.18 | 1.31 | 5.00 | 10.99 | 13.15 | 14.54 |
| Peer[b] | 4.50 | .92 | 4.92 | .79 | 4.34 | .62 | 3.00 | 1.21 | 2.31 | 5.20 | 9.41 | 10.66 |
| Supervisor[b] | 4.34 | .97 | 4.88 | 1.13 | 3.80 | .99 | 3.72 | 1.20 | 4.81 | 7.76 | 10.78 | 8.57 |
| Total | 4.42 | .93 | 5.02 | .90 | 4.13 | .93 | 3.30 | 1.27 | 4.05 | 8.36 | 11.13 | 11.55 |
| Total robot partners | 4.50 | .86 | 4.83 | 1.05 | 4.31 | .92 | 3.34 | 1.34 | 3.52 | 6.91 | 10.71 | 10.94 |

[a] $n = 33$. [b] $n = 32$.

ported lower levels of personal responsibility ($M = 4.64$ vs. $5.02$). An ANOVA contrasting these two conditions confirms that the effect is significant, $F(2, 189) = 6.37$, $p = .01$, even when mood is included in the analysis, $F(2, 187) = 5.29$, $p = .02$. Our measure of attribution of credit shows a similar pattern with less credit being attributed to the robot in the machine-like as compared with the human-like conditions ($M = 4.13$ vs. $4.50$). An ANOVA contrasting the human-like and machine-like robot conditions suggests that the effect on attribution of credit, $F(2, 189) = 8.41$, $p = .004$, is significantly different in the two conditions even when gender is included as a covariate, $F(2, 188) = 8.35$, $p = .004$. There was not, however, a significant difference when predicting attribution of blame, $F(2, 188) = .24$, $p = .63$. Therefore, although little support is provided for Hypothesis 1a, substantial support is provided for Hypothesis 1b in which we argue that robots with a human-like as compared with a machine-like appearance will reduce the extent to which people feel responsible for the task.

In an additional analysis of the effect of appearance, we estimated linear regression equations including the human confederate (baseline condition) as part of the human-likeness scale (machine-like robot, human-like robot, human confederate). Doing so allowed us to test the reasoning that people would rely more on partners and share more responsibility for the task when the partners were more human-like (including being human). Although no strong linear effect was found for responsibility, there was a strong positive relation with reliance ($\beta = .16$, $p = .007$) and attribution of credit ($\beta = .17$, $p = .004$), indicating that people exhibited more reliance and attributed more credit to their partners as their partners displayed more human-like (or human) characteristics.

In the logical arguments leading to our hypotheses, we reasoned that people will feel a stronger sense of social identity with human-like robots than they will with machine-like robots, and that shared identity might contribute to more reliance and shared responsibility. From the videotapes, we coded individualistic language (e.g., *I, me, you, your*) and collective language (e.g., *us, we, our*) as a means of measuring the extent to which the participants were expressing a sense of shared social identity with the robot. Although not statistically significant, participants used less individualistic language with human-like robots than with machine-like robots ($M = 10.23$ vs. $11.13$), $F(2, 177) = .31$, $p = .58$.[3] However, they also used less collective language ($M = 2.93$ vs. $4.05$), $F(1,$

---

3. We also conducted the analyses including the two outliers (in excess of 100 uses of individualistic language). Both of the outliers were in the human-like robot supervisor conditions. When included, the mean for the human-like robot supervisor condition is $17.76$ ($SD = 35.04$), and the mean for the human-like robot conditions is $13.18$ ($SD = 21.99$). When we include the outliers in the analysis of variance, the difference between the human-like and machine-like robot conditions is not statistically significant, $F(2, 179) = .68$, $p = .41$.

179) $= 1.42$, $p = .25$. These data provide conflicting results and, thus, little support for our arguments about shared identity being at the root of differences in reliance and responsibility when working with human-like versus machine-like robots. Normalizing the data to get the fraction of all pronouns used that were collective or individualistic produced no statistically significant effects. The pattern of these data suggests that people were more talkative overall with machine-like robots as compared with human-like robots. The data could be interpreted as support for a common ground explanation. That is, when people perceive less common ground between themselves and a partner, they tend to talk more (Fussell & Krauss, 1992). They talk more, in part, to provide the partner with the background required to interpret future interactions and, in part, to gather more information about what the partner knows. To evaluate this possibility further, we coded the videotapes for the number of factual questions that the participant asked of the confederate. Consistent with a common-ground explanation, participants who worked with the human-like robot asked fewer factual questions on average $(M = 2.58)$ than those who worked with the machine-like robots $(M = 3.45)$, although this difference was not significant, $F(1, 182) = 1.29$, $p = .26$.

## 4.2. Effects of Status

In Hypotheses 2a and 2b, we argued for a main effect of status. We posited that people will rely more on a robotic partner and feel less responsible for the task when the partner is assigned a supervisory as compared with a subordinate or peer position relative to the participant. We found little support for Hypothesis 2a. People relied more on the robot peers than they did robot subordinates or supervisors $(M = 4.41$ supervisors vs. 4.60 peers vs. 4.51 subordinates), and the difference between the supervisor and the other status conditions was not significant, $F(1, 182) = 1.13$, $p = .29$. Analyzing Hypothesis 2b, participants reported feeling less responsible when collaborating with a robot supervisor as compared with a robot peer or subordinate $(M = 4.74, 4.80, 4.94$, respectively), but participants also reported that less credit was due to the partner when it was a supervisor, which is the opposite of what we had hypothesized $(M = 4.08, 4.39, 4.47$, respectively). Although the effect for responsibility was not significant, $F(1, 191) = .66$, $p = .42$, when conducting two-way ANOVAs contrasting the supervisor condition with the other status conditions, the effect of status on attribution of credit was significant, $F(1, 191) = 6.73$, $p = .01$. That is, participants attributed significantly less credit to the robot supervisor as compared with the robot peer and subordinate. Paradoxically, we also found that participants were more likely to blame robot supervisors as compared with robot peers and subordinates for errors and mistakes that were made $(M = 3.83, 3.07, 3.13$, respectively) and that this difference was highly significant, $F(1, 191) = 13.53$, $p < .001$. Overall, it seems that participants were much more critical of the robot in a supervisory position.

As with our Hypotheses 1a and 1b, we had an underlying linear assumption in our status variable, suggesting that status would increase the extent to which people relied on the partner and would decrease their sense of responsibility for the task. Therefore, we conducted linear regressions with status as a scale (supervisor, peer, subordinate) and human likeness as a variable. The only significant linear effects were found when predicting attribution of credit ($\beta = -.137$, $p = .001$) and attribution of blame ($\beta = .21$, $p = .003$), suggesting that as the status of the robot target increases, people attribute less credit for successes and more blame for failures in performance.

## 4.3. Interaction Between Human Likeness and Status

Our final hypothesis predicted an interaction effect between human likeness of the robots and the robot's status. In Hypothesis 3, we argued that people will feel most responsibility for the task when they work with a machine-like robot subordinate. To test this hypothesis, we conducted a two-way ANOVA analysis with only the robot conditions included in the analysis. We found a significant effect in the expected direction for responsibility when contrasting the machine-like subordinate conditions with all other robot conditions, $F(1, 193) = 6.37$, $p = .01$. We found little effect for attribution of credit, $F(1, 193) = .26$, $p = .61$; or attribution of blame, $F(1, 193) = .58$, $p = .45$. These analyses provide some evidence that people will feel most responsible for the task when they are collaborating with a machine-like robot that is presented as a subordinate. A similar test predicting reliance showed no significant effect, as we expected, $F(1, 184) = .36$, $p = .55$.

A summary of the support found for each of our hypotheses is detailed in Figure 6.

## 5. DISCUSSION

As far as we know, ours is one of the first systematic, controlled experiments comparing people's responses to human coworkers versus robot coworkers; and to more humanoid versus less humanoid robots. Our findings suggest that there are significant differences in the extent to which people will rely on robots as compared with human work partners. When working with a person instead of a robot, participants relied more on the partner's advice and were less likely to ignore their counsel. We found, however, only marginal support for the idea that people would feel less burden of responsibility when interacting with another person as compared with a robot. It appears from these data that

*Figure 6.* Summary of hypotheses and results.

| Hypotheses | Results |
| --- | --- |
| Human-like versus machine-like robots | |
| Hypothesis 1a: People will rely on a human-like robot partner more than on a machine-like robot partner. | Not supported |
| Hypothesis 1b: People will feel less responsible for the task when collaborating with a human-like robot partner than with a machine-like robot partner. | Supported |
| Relative status of robot coworkers | |
| Hypothesis 2a: People will rely on the robot partner more when it is characterized as a supervisor than when it is characterized as a subordinate or peer. | Not supported |
| Hypothesis 2b: People will feel less responsible for the task when collaborating with a robot partner who is a supervisor than with a robot partner who is a subordinate or peer. | Mixed support |
| Interactions between human-likeness and status | |
| Hypothesis 3: People will feel the greatest amount of responsibility when collaborating with machine-like robot subordinates as compared with machine-like robot peers and supervisors; and as compared with human-like robot subordinates, peers, and supervisors. | Mixed support |

participants collaborated more with the human partners than with the robot partners but still did not necessarily cede responsibility to them.

Comparing robots with different appearances, our data show that interacting with a more machine-like robot may increase the personal responsibility that people feel for the task. This effect was increased when status was added to the manipulation. Our data indicate that participants felt most responsible when interacting with the machine-like subordinate, suggesting that a machine-like appearance coupled with the framing of a subordinate position may result in the highest levels of human responsibility. Knowing this may be useful when it is important for workers to feel the full weight of responsibility for the task in which they are engaged. When people feel more responsibility for the task, mishaps may be avoided because people explore more options and are more diligent about finding an appropriate solution (see Roberts et al., 1994). When people feel responsibility for the task, they may also be less likely to trust the robotic partner to perform tasks for which the robot is ill equipped or when the robot becomes ill equipped due to unanticipated changes to the task requirements or the environment. On the other hand, our results suggest that humanoid robots may be appropriate for situations in which the burden of responsibility can or should be attenuated for the people involved in the task.

For example, humanoid robots may be preferred when human complacency is not a concern or when the consequences of a risky task would be too difficult for a human to bear.

We hypothesized, as have others (e.g., Parise et al., 1996), that one of the reasons that people may respond differently to human-like technologies than to machine-like technologies is because they feel more similar to the former and thus experience more shared identity with them. Our behavioral measures of shared identity, however, provided little evidence to this effect. We did not see a difference in the individualistic or collective language used by participants in the different conditions. On the other hand, participants in the machine-like robot conditions appeared to talk more with the robot than did participants in the human-like robot conditions, suggesting that they might have perceived less common ground with these robots, and felt they had to explain themselves more, or provide more instruction.

Our status manipulation generated mixed effects. When collaborating with supervisors, participants attributed less credit and more blame to their partner. This effect suggests that, as in the popular Dilbert cartoon (see www.dilbert.com), the supervisor was viewed as undeserving and was blamed for most of the problems encountered on the task. Additional research is needed to assess the robustness of this effect. It is possible that the effect in our study is specific to the task being performed and to the role we assigned to the supervisor. The task was relatively straightforward, albeit with some ambiguity, and did not allow the supervisor to display a particularly impressive level of skill or competence. Our results may suggest that when people or robots are put in supervisory positions without commensurate skills or authority, their subordinates will respond negatively. A situation in which the partner has substantially more skill or knowledge relative to the participant might reveal somewhat different effects. It is also possible, however, that the comparatively young student population in our sample have beliefs that are consistent with the "Dilbert effect," maintaining fairly negative impressions of those in supervisory roles.

This study is an early foray into the examination of people's responses to professional service robots. As such, many questions remain, and additional studies are needed to fully understand how people will respond to and work with professional service robots on collaborative tasks. One area that merits exploration, for example, is the nature of the task being performed. The task we utilized here was a relatively mundane parts-sorting task that required movement and knowledge on the part of the robot. Even in the ambiguous parts of the task, participants demonstrated high levels of reliance on their partner ($M = 4.58$ out of 5). We expect that reliance was high because the task was routine and posed little risk, and the robot was clearly equipped to perform the role assigned to it. Over time, robots are likely to assume tasks that are substantially more complex, risky, and uncertain than this experimental task (e.g., Burke, Murphy, Coovert, & Riddle, 2004). Situations in which people working with a robot are already cognitively overloaded and do not have the capacity to monitor the robot's actions are also likely. For example, search and rescue robots may work in tandem with rescue workers in extreme weather conditions, such as in the aftermath of a hurricane. In such situations, people are overworked and experience stress, uncertainty, and physical danger. Based on the research reported here, we anticipate that humanoid robots may be appropriate for tasks that are complex or risky because people will more readily delegate responsibility to them. Although our task did not present high risk to the participants, our participants reported being reasonably well motivated to participate ($M = 6.97$ on a 10-point scale); we anticipate that stronger motivation and higher risk might strengthen these results. Substantial research, however, is needed to fully understand the interplay between the design of the robot, the task being performed, the interaction between the person and the robot's skill and knowledge, the amount of perceived risk, and people's willingness to rely on the robot.

Although the focus of our research was on robots in the work environment, this study was conducted in a controlled laboratory setting. Doing so enabled us to maintain control over the conditions being tested. At the same time, realistic aspects of the work environment were intentionally designed out. For example, participants in our study worked in dyads and did not interact with other coworkers or friends. We believe that people's responses to robots in the work environment will be significantly influenced by the social and organizational context in which they are embedded (see Siino & Hinds, 2004). Robots also could have a significant and unanticipated effect on the dynamics of work teams. Existing research suggests that the nature of effective team processes may be different when automated systems are introduced (see Bowers, Oser, Salas, & Cannon-Bowers, 1996). It will be fruitful to investigate the effect of human-like and machine-like robots on the dynamics of teams and organizations.

In examining human likeness versus machine likeness of the robot, we chose to create robots that were a composite of a variety of human-like and machine-like features. The human-like robot had facial features, a torso, arms, and legs. It had the appearance of a man clothed in a denim shirt, khaki pants, and a baseball cap. The machine-like robot was metallic and angular with none of the physical human-like features previously described. We created these two conditions to make an initial determination of the impact of human likeness in a robot partner. We believe, however, that it will be important to decompose and examine the independent effects of features such as eyes, mouth, and legs. One also could look individually at features that we held constant: a human-like voice, the absence of human-like gestures, and the robot's movements (e.g., rolling vs. walking). Along with others (e.g., Jensen, Farnham, Drucker, & Kollock, 2000; Nass & Lee, 2000), we think it quite likely that each of these features may have some effect on perceived human likeness and on perceptions of the robot coworker. Given the power of voice, it is possible that the effects that we found could be duplicated by manipulating only the voice (natural human voice vs. machine-generated voice). Similarly, behavioral manipulations may provide a powerful way to convey human likeness in robots (see Breazeal, 1999) and may generate similar effects. Research into each of these dimensions would be a contribution to this field of work.

It is also important to note that although we manipulated human likeness of the robot, the robot we used was nowhere near as human like as advanced robots are and can be. For example, Honda's ASIMO has a more humanoid form than we were able to produce in this study. In our pretest, our machine-like robot was rated on average 2.90, and our human-like robot on average 3.69, on a 7-point scale of human likeness. It is possible that our effects for reliance are weak because our human-like robot was not extremely human like and the task did not require advanced human-like behaviors. We anticipate that future research examining human reliance using increasingly humanoid robotic forms and behaviors will find stronger effects. This work also will help us understand the so-called uncanny valley—a space in which robots evoke expectations of human likeness but are not quite human and, therefore, create discomfort (see DiSalvo et al., 2002; Reichard, 1978).

Other factors that we anticipate will have a significant effect on human–robot interaction are the expectations that people have of robots and the experience they have with them. As people gain experience, the novelty of the technology wears off and people develop ways to adapt the technology to better fit their needs (e.g., Barley, 1986; Orlikowski, 2000). In this study, having some experience with robots (e.g., a class or two) did not affect our results. Having taken a class or two about robotics, however, may not be enough to establish clear expectations about the capabilities of robots in general or about a specific robot in particular. As people gain more experience with robots and with the particular robot with which they are working, we expect they will develop new mental models of the robots' capabilities, revise their perceptions of how the robot fits into the work environment, and make alterations to the robot or their use of it to better accommodate their needs. This gain in experience will no doubt affect people's collaboration with the robot. Ideally, longitudinal studies will help to inform these questions.

Although there is substantial research required to fully understand human–robot interaction on collaborative tasks, the research reported here provides some initial findings that suggest how people may respond to humanoid professional service robots and the conditions under which humanoid robots may be preferable to machine-like robots. These findings provide input to guide the decisions of those designing robots and determining the roles that robots will play in the work setting.

More broadly, we view this research as an early effort using the theoretical and empirical foundations of social psychology and organizational behavior to inform the design of robots for the work environment. We believe that continued work in this area and ongoing collaboration between social scientists and roboticists to identify and explore questions of importance can make a significant contribution to the field of robotics and improve the likelihood for successful implementation of professional service robots.

## NOTES

*Acknowledgments.* We are grateful for the helpful suggestions made by Sara Kiesler and three anonymous reviewers on earlier versions of this article. We also appreciate the assistance provided by André Guérin and Marissa Rolland in conducting the experiment. Finally, we thank Breyana Rouzan, Kimberly Roberts, Clint Martin, and Catherine Liang for their help coding the videotapes.

*Support.* This study was funded by NSF Grant IIS-0121426 to Pamela J. Hinds.

*Authors' Present Addresses.* Pamela J. Hinds, Management Science & Engineering, Terman 424, Stanford University, Stanford, CA 94305–4026. E-mail: **phinds@stanford.edu**. Teresa L. Roberts, PeopleSoft, Inc., 4460 Hacienda Drive, Pleasanton, CA 94588–8618. E-mail: **terry_roberts@peoplesoft.com**. Hank Jones, Aerospace Robotics Laboratory, Durand 250, Stanford University, Stanford, CA 94305. E-mail: **hlj@arl.stanford.edu**.

*HCI Editorial Record.* First manuscript received December 5, 2002. Revision received June 1, 2003. Accepted by Sara Kiesler. Final manuscript received September 14, 2003. — *Editor*

## REFERENCES

Ambrose, R. O., Askew S. R., Bluethmann, W., & Diftler, M. A. (2001). Humanoids designed to do work. *Proceedings of the IEEE 2001 International Conference on Humanoid Robots, Robotics and Automation Society.* Tokyo, Japan.

Barley, S. (1986). Technology as an occasion for structuring: Evidence from observation of CT scanners and the social order of radiology departments. *Administrative Science Quarterly, 31,* 78–108.

Bowers, C. A., Oser, R. L, Salas, E., & Cannon-Bowers, J. A. (1996). Team performance in automated systems. In R. Parasuraman & M. Mouloua (Eds.), *Automation and human performance: Theory and applications* (pp. 243–265). Mahwah, NJ: Lawrence Erlbaum Associates, Inc.

Breazeal, C. (1999). Robot in society: Friend or appliance? *Proceedings of the Agents 99 Workshop on Emotion-Based Agent Architectures.* Seattle, Washington.

Breazeal, C., & Scassellati, B. (1999). *How to build robots that make friends and influence people*. Paper presented at the 1999 IEEE/RSJ International Conference on Intelligent Robots and Systems (IROS–99), Kyonjiu, Korea.

Brooks, R. (2002). Humanoid robots. *Communications of the ACM, 45*(3), 33–38.

Brooks, R., & O'Reilly, U.-M. (2002). *Humanoid robotics group, MIT artificial intelligence laboratory.* Retrieved November 11, 2002, from http://www.ai.mit.edu/projects/humanoid-robotics-group/

Burke, J. L., Murphy, R. R., Coovert, M. D., & Riddle, D. L. (2004). Moonlight in Miami: A field study of human–robot interaction in the context of an urban search and rescue disaster response training exercise. *Human–Computer Interaction, 19,* 85–116. [this special issue]

Dautenhahn, K. (1999). Robots as social actors: AURORA and the case of autism. *Proceedings of the Third International Cognitive Technology Conference.* San Francisco, CA.

DiSalvo, C. F., Gemperle, F., Forlizzi, J., & Kiesler, S. (2002). All robots are not created equal: The design and perception of humanoid robot heads. *Proceedings of the DIS 2002 Conference on Designing Interactive Systems.* New York: ACM.

Fong, T., Nourbakhsh, I., & Dautenhahn, K. (2002). *A survey of social robots* (CMU Robotics Institute Technical Report CMU–RI–TR–02–29). Pittsburgh, PA.

Forgas, J. P., Bower, G. H., & Moylan, S. J. (1990). Praise or blame? Affective influences on attributions for achievement. *Journal of Personality and Social Psychology, 59,* 809–819.

Fussell, S., & Krauss, R. M. (1992). Coordination of knowledge in communication: Effects of speakers' assumptions about what others know. *Journal of Personality and Social Psychology, 62,* 378–391.

Gawande, A. (2002). *Complications: A surgeon's notes on an imperfect science.* New York: Holt.

Goetz, J., Kiesler, S., & Powers, A. (2003). Matching robot appearance and behavior to tasks to improve human–robot cooperation. *Proceedings of the 12th IEEE Workshop on Robot and Human Interactive Communication. RO-MAN 2003.* San Francisco, CA.

Goodnow, J. J. (1996). Collaborative rules: From shares of the work to rights to the story. In P. Baltes & U. Staudinger (Eds.), *Interactive minds* (pp. 163–193). Cambridge, UK: Cambridge University Press.

Grabowski, M., & Roberts, K. (1997). Risk mitigation in large-scale systems: Lessons from high reliability organizations. *California Management Review, 39,* 152–162.

Hashimoto, S., Narita, S., Sugano, S., Takanishi, A., Shirai, K., & Kobayashi, T. (2002). *History of humanoid robot at Waseda University* (Humanoid Robotics Institute, Waseda University, Japan). Retrieved November 11, 2002, from http://www.humanoid.waseda.ac.jp/history.html

Isaacs, E., & Clark, H. (1987). References in conversation between experts and novices. *Journal of Experimental Psychology: General, 116,* 26–37.

Ishiguro, K. (2003). *Intelligent robotics laboratory, Osaka University.* Retrieved July 28, 2003, from http://www.ed.ams.eng.osaka-u.ac.jp/

Jensen, C., Farnham, S. D., Drucker, S. M., & Kollock, P. (2000). The effect of communication modality on cooperation in online environments. *Proceedings of the CHI 2002 Conference on Human Factors in Computer Systems.* New York: ACM.

Kiesler, S., & Goetz, J. (2002). Mental models of robotic assistants. *Proceedings of the CHI 2002 Conference on Human Factors in Computer Systems.* New York: ACM.

Kraut, R., Fussell, S., Lerch, F., & Espinosa, A. (2002). *Coordination in teams: Evidence from a simulated management game* (Working Paper). Retrieved May 26, 2003, from http://www.andrew.cmu.edu/user/sfussell/Manuscripts_Folder/Team_Co-ordination.pdf

Nass, C., Fogg, B. J., & Moon, Y. (1996). Can computers be teammates? *International Journal of Human–Computer Studies, 45,* 669–678.

Nass, C., & Lee, K. M. (2000). Does computer-generated speech manifest personality? An experimental test of similarity-attraction. *Proceedings of the CHI 2000 Conference on Human Factors in Computer Systems.* New York: ACM.

Nass, C., Steuer, J., Tauber, E., & Reeder, H. (1993). Anthropomorphism, agency, and ethopoeia: Computers as social actors. *Proceedings of the INTERCHI 93 Conference on Human Factors in Computer Systems.* New York: ACM.

Okie, S. (2002, April 3). Robots make the rounds to ease hospitals' costs. *The Washington Post,* p. A03.

Olson, G. M., & Olson, J. S. (2000). Distance matters. *Human–Computer Interaction, 15,* 139–179.

Orlikowski, W. (2000). Using technology and constituting structures: A practice lens for study technology in organizations. *Organization Science, 11,* 404–428.

Parise, S., Kiesler, S., Sproull, L., & Waters, K. (1996). My partner is a real dog: Cooperation with social agents. *Proceedings of the CSCW 96 Conference on Computer-Supported Cooperative Work.* New York: ACM.

Pfeffer, J., Cialdini, R., Hanna, B., & Knopoff, K. (1998). Faith in supervision and the self-enhancement bias: Two psychological reasons why managers don't empower workers. *Basic and Applied Social Psychology, 20,* 313–321.

Reeves, B., & Nass, C. (1996) *The media equation.* Cambridge, UK: Cambridge University Press.

Reichard, J. (1978). *Robots: Fact, fiction, and prediction.* New York: Penguin.

Roberts, K. H., Stout, S. K., & Halpern, J. J. (1994). Decision dynamics in two high reliability military organizations. *Management Science, 40,* 614–624.

Sande, G. N., Ellard, J. H., & Ross, M. (1986). Effect of arbitrarily assigned labels on self-perceptions and social perceptions: The mere position effect. *Journal of Personality and Social Psychology, 50,* 684–689.

Scholl, B. J., & Tremoulet, P. (2000). Perceptual causality and animacy. *Trends in Cognitive Science, 4,* 299–309.

Sheridan, T. B. (1992). *Telerobotics, automation, and human supervisory control.* Cambridge, MA: MIT Press.

Siino, R., & Hinds, P. (2004). *Making sense of new technology: Organizational sensemaking of autonomous mobile robots as a lead-in to structuring.* Unpublished manuscript, Stanford University.

Simmons, R., & Nourbakhsh, I. (2002). *Social robots project, Carnegie Mellon University.* Retrieved November 11, 2002, from http://www-2.cs.cmu.edu/~social/

Sparaco, P. (1994). A330 crash to spur changes at Airbus. *Aviation Week and Space Technology, 141*(6), 20–22.

Squeo, A. (2001, December 13). Meet the newest recruits: Robots. *The Wall Street Journal*, p. B01.

Steuer, J. (1995). *Vividness and source of evaluation as determinants of social responses toward mediated representations of agency.* Unpublished doctoral dissertation, Stanford University, California.

Strodtbeck, F. L., James, R. M., & Hawkins, D. (1957). Social status in jury deliberations. *American Journal of Sociology, 22,* 713–719.

Swim, J. K., & Sanna, L. J. (1996). He's skilled, she's lucky: A meta-analysis of observers' attributions for women's and men's successes and failures. *Personality and Social Psychology Bulletin, 22,* 507–519.

Tetlock, P. E. (1985). Accountability: The neglected social context of judgment and choice. In L. L. Cummings & B. M. Staw (Eds.), *Research in organizational behavior* (Vol. 7, pp. 297–332). Greenwich, CT: JAI.

Thrun, S. (2004). Toward a framework for human–robot interaction. *Human–Computer Interaction, 19,* 9–24. [this special issue]

Turkle, S. (1984). *The second self: Computers and the human spirit.* New York: Simon & Schuster.

United Nations and The International Federation of Robotics. (2002). *World Robotics 2002.* New York: United Nations.

Wall, T. D., Jackson, P. R., & Davids, K. (1992). Operator work design and robotics system performance: A serendipitous field study. *Journal of Applied Psychology, 77,* 353–362.

# 2004 SUBSCRIPTION ORDER FORM

Please ☐ enter ☐ renew my subscription to:

## HUMAN–COMPUTER INTERACTION

A JOURNAL OF THEORETICAL, EMPIRICAL, AND METHODOLOGICAL ISSUES OF USER SCIENCE AND OF SYSTEM DESIGN

### Volume 19, 2004, Quarterly — ISSN 0737–0024/Online ISSN 1532–7051

#### SUBSCRIPTION PRICES PER VOLUME:

| Category: | Access Type: | Price: US-Canada/All Other Countries |
|---|---|---|
| ☐ Individual | Online & Print | $60.00/$90.00 |

Subscriptions are entered on a calendar-year basis only and must be paid in advance in U.S. currency—check, credit card, or money order. Prices for subscriptions include postage and handling. **Journal prices expire 12/31/04.** NOTE: Institutions must pay institutional rates. Individual subscription orders are welcome if prepaid by credit card or personal check. **Please note:** A $20.00 penalty will be charged against customers providing checks that must be returned for payment. This assessment will be made only in instances when problems in collecting funds are directly attributable to customer error.

☐　**Check Enclosed** (U.S. Currency Only)　　　**Total Amount Enclosed $**_____

☐　**Charge My**: ☐ VISA ☐ MasterCard ☐ AMEX ☐ Discover

Card Number _____　Exp. Date_____/_____

Signature_____

*(Credit card orders cannot be processed without your signature.)*
**PRINT CLEARLY** for proper delivery. STREET ADDRESS/SUITE/ROOM # REQUIRED FOR DELIVERY.

Name_____

Address_____

City/State/Zip+4_____

Daytime Phone #_____E-mail address_____
*Prices are subject to change without notice.*

**For information about online subscriptions, visit our website at** *www.leaonline.com*

Mail orders to: **Lawrence Erlbaum Associates, Inc.,** Journal Subscription Department
10 Industrial Avenue, Mahwah, NJ 07430; **(201) 258–2200; FAX (201) 760–3735; journals@erlbaum.com**

## LIBRARY RECOMMENDATION FORM

*Detach and forward to your librarian.*

☐ I have reviewed the description of the *Human-Computer Interaction* and would like to recommend it for acquisition.

## HUMAN–COMPUTER INTERACTION

A JOURNAL OF THEORETICAL, EMPIRICAL, AND METHODOLOGICAL ISSUES OF USER SCIENCE AND OF SYSTEM DESIGN

### Volume 19, 2004, Quarterly — ISSN 0737–0024/Online ISSN 1532–7051

| Category: | Access Type: | Price: US-Canada/All Other Countries |
|---|---|---|
| ☐ Institutional | Online & Print | $475.00/$505.00 |
| ☐ Institutional | Online Only | $405.00/$405.00 |
| ☐ Institutional | Print Only | $430.00/$460.00 |

Name_____Title _____

Institution/Department_____

Address_____

E-mail Address_____

*Librarians, please send your orders directly to LEA or contact from your subscription agent.*

**Lawrence Erlbaum Associates, Inc.,** Journal Subscription Department
10 Industrial Avenue, Mahwah, NJ 07430; **(201) 258–2200; FAX (201) 760–3735; journals@erlbaum.com**